Jens Burgert

On direct and inverse problems related to longitudinal impact of non-uniform elastic rods

Karlsruher Institut für Technologie
Schriftenreihe des Instituts für Technische Mechanik

Band 36

Eine Übersicht aller bisher in dieser Schriftenreihe erschienenen Bände finden Sie am Ende des Buchs.

On direct and inverse problems related to longitudinal impact of non-uniform elastic rods

by
Jens Burgert

Scientific
Publishing

Dissertation, Karlsruher Institut für Technologie
KIT-Fakultät für Maschinenbau

Tag der mündlichen Prüfung: 07. Dezember 2020
Hauptreferent: Prof. Dr.-Ing. Wolfgang Seemann
Korreferent: Prof. Dr. Dr. h.c. Peter Hagedorn

Impressum

KIT Scientific Publishing

Karlsruher Institut für Technologie (KIT)
KIT Scientific Publishing
Straße am Forum 2
D-76131 Karlsruhe

KIT Scientific Publishing is a registered trademark
of Karlsruhe Institute of Technology.
Reprint using the book cover is not allowed.

www.ksp.kit.edu

Print on Demand 2021 – Gedruckt auf FSC-zertifiziertem Papier

ISSN 1614-3914
ISBN 978-3-7315-1087-1
DOI 10.5445/KSP/1000129237

Danksagung

Die vorliegende Arbeit entstand während meiner Tätigkeit als wissenschaftlicher Mitarbeiter am Institut für Technische Mechanik (ITM), Abteilung Dynamik/Mechatronik des Karlsruher Instituts für Technologie (KIT).

An erster Stelle gilt mein Dank Herrn Prof. Dr.-Ing. Wolfgang Seemann für die fachliche Betreuung. Ich danke ihm für die vertrauensvolle Zusammenarbeit, die wissenschaftliche Freiheit und die stete Unterstützung und Förderung. Hervorheben möchte ich seine Unterstützung beim Aufbau internationaler Kooperationen mit anderen Universitäten, von denen ich fachlich und persönlich sehr profitiert habe.

Weiterhin bedanke ich mich bei Prof. Dr. Dr. h.c. Peter Hagedorn für die Übernahme des Korreferats und den damit verbundenen Anregungen zu meiner Arbeit. Herrn Prof. Dr. mont. Christoph Kirchlechner danke ich für die Übernahme des Prüfungsvorsitzes.

Ebenfalls danke ich Herrn Prof. Dr.-Ing. habil. Alexander Fidlin für die stets ehrlichen Fragen und Anmerkungen in den Seminaren. Herrn Prof. Dr.-Ing. Carsten Proppe danke ich für die angenehme Zusammenarbeit am Institut.

Den emeritierten Professoren Prof. Dr.-Ing. Dr. h.c. Jörg Wauer, Prof. Dr.-Ing. Walter Wedig und Prof. Dr.-Ing. Dr. h.c. Jens Wittenburg danke ich für viele interessante Gespräche. Ihre Teilnahme am Institutsleben ist eine große Bereicherung. Ich hoffe, dass Sie dem wissenschaftlichen Nachwuchs noch viele Jahre zur Seite stehen.

Die Forschungsaufenthalte an der Universität Tampere in Finnland und der Universität Uppsala in Schweden haben der vorliegenden Arbeit entscheidende Impulse gegeben. Großen Anteil am Gelingen der Aufenthalte haben die Professoren Prof. Dr. Erno Keskinen und Prof. Dr. Bengt Lundberg. Ihnen danke ich von Herzen.

Des Weiteren bedanke mich bei allen Kolleginnen und Kollegen für die wundervollen Jahre am Institut. Neben den fachlichen Diskussionen werden mir vor allem das kollegiale Miteinander, die Hilfsbereitschaft und die persönlichen Gespräche in positiver Erinnerung bleiben. Hervorheben möchte ich meinen Zimmerkollegen Jimmy Aramendiz, mit dem ich über jedes Thema philosophieren konnte und mit dem ich auch abseits der Tätigkeit am Institut viele schöne Stunden verbringen durfte.

Ebenfalls danke ich allen Abschlussarbeitern und Hilfswissenschaftlern, deren Betreuung mir sehr viel Freude bereitet hat und von deren Arbeit und Fragen ich sehr profitiert habe.

Ganz besonders möchte ich mich bei meiner Familie bedanken. Ihr konstanter Rückhalt und ihre Unterstützung in jeglicher Hinsicht haben mich tief beeindruckt und berührt. Durch sie habe ich nie aus den Augen verloren, was die wirklich wichtigen Dinge im Leben sind. Dabei möchte ich besonders meinen Eltern für ihre Hingabe und ihre bedingungslose Liebe danken.

Zu guter Letzt gilt mein größter Dank meiner lieben Frau Kristin, die mich stets motiviert und unterstützt hat, insbesondere in dem turbulenten letzten Jahr. Ich danke ihr vor allem dafür, dass sie mein Leben auf eine so wundervolle Art und Weise bereichert.

Karlsruhe, den 10. Dezember 2020
Jens Burgert

Zusammenfassung

Das Ziel der Arbeit ist die Entwicklung einer Methode zur Untersuchung inverser Probleme in aufeinandertreffenden Stäben. Dabei soll die Impedanz des auftreffenden Stabs ermittelt werden, sodass eine vorgegebene Stoßkraft erzeugt wird (*Impedanzproblem*). Die Methode ist für viele experimentelle und praktische Anwendungen relevant. Beim Schlagbohrverfahren stellt sich beispielsweise die Frage, wie die Impedanz des auftreffenden Stabs gewählt werden kann, sodass die einlaufende Kraftwelle bezüglich der Effizienz optimiert wird.

Das *Impedanzproblem* wird nur von wenigen Forschern behandelt. Der vielversprechendste Ansatz löst das inverse Problem mithilfe einer Integralgleichung. Dabei werden mehrere Näherungsmethoden angewandt, um die Impedanz des auftreffenden Stabs zu ermitteln. Im Rahmen dieser Arbeit wird eine Methode entwickelt, die auf der Idee differentieller Methoden basiert. Die Methode nutzt die Struktur des Wellenausbreitungsproblems aus. Hierfür werden die Stäbe durch Elemente mit stückweiser konstanter Impedanz diskretisiert. Zunächst wird die Impedanz des ersten Elements ermittelt, welches direkt auf die gestoßene Stange auftrifft. Alle weiteren Elemente beeinflussen die Stoßkraft zeitverzögert. Je weiter die Elemente von der Stoßfläche entfernt sind, desto später ist ihr Einfluss sichtbar. Zum Zeitpunkt des erstmaligen Einflusses des jeweiligen Elements wird die zugehörige Impedanz so gewählt, dass die vorgegebene Stoßkraft erreicht wird. Dadurch werden die diskretisierten Impedanzen des auftreffenden Stabs rekursiv bestimmt.

Für stückweise konstante Impedanzfunktionen liefert die entwickelte Methode exakte Lösungen in geschlossener Form. Des Weiteren wird eine Bedingung hergeleitet, die angibt, ob eine Lösung zu dem inversen Problem existiert. Es wird gezeigt, dass jedes zusätzliche Element, dessen Impedanz sich von der Impedanz des vorigen Elements unterscheidet, die Wahrscheinlichkeit der Existenz einer Lösung verringert.

Das zugrundeliegende 1D-Modell wird experimentell validiert, indem die Simulationsergebnisse mit Messungen verglichen werden. Es wird insbesondere gezeigt, dass das Modell auch für komplexe Geometrievariationen der Stäbe gültig ist.

Abstract

Impacting rods are used in many practical and experimental applications. The analysis problem of finding the time-dependent impact force for given properties of the impacting rods and known initial velocities is thoroughly studied in literature. However, very little literature has been published which addresses its inverse synthesis problem: Find the location-dependent impedance function of the impacting rod that generates a prescribed impact force.

In this contribution, a method that solves the synthesis problem is developed. The impedances of the linear elastic rods are discretized by elements of piecewise constant impedance. This enables making use of the analytical solution which exists for these elements. The method utilizes that each element of the unknown impactor influences the prescribed force at different time instants for the first time. At this specific time, the impedance of the corresponding element is adjusted so that the actual force matches the prescribed force. This procedure is repeated recursively until the impedances of all impactor elements are determined. For piecewise constant impedance functions, the developed method delivers exact solutions in closed-form. Moreover, a condition is derived which states if a physically meaningful solution exists for the considered element. Thus, it is known if and how long a solution of the inverse problem exists. The working principle of the method is demonstrated with several examples, including the application in percussion drilling.

Finally, the underlying 1D model is validated experimentally. To this end, a single-hit test rig is set up. The comparison of simulation versus experimental results yields good accordance both for standard and complex geometries of the impacting rods.

Contents

1 Introduction

Wave propagation phenomena are present in anyone's daily routine. Every time we talk to someone, pressure waves are generated which propagate through the air to the ears of the listener. The sound is created by inhaled air which is sent back out of the mouth. On the way from the lung to the mouth, the airflow is shaped by the vocal cords (excitation) and the cross sectional area variation of the vocal tract. Humans intuitively choose the appropriate excitation and cross section area of the vocal tract while talking. However, it is of interest to identify which vocal tract cross section area leads to the corresponding sound. The direct problem of determining the pressure at the lips for given excitation and cross section area of the vocal tract is very well understood. As opposed to this, the inverse problem of finding the vocal tract's cross section area for given excitation and pressure measurement at the lips is much more demanding.

Typical questions that arise when dealing with inverse problems are:

1. Does a solution of the inverse problem exist?
2. Is the solution unique?

Another example of an inverse wave propagation problem is the determination of the characteristic rod impedance function by applying a force impulse at one end of the rod and measuring the impulse response. Similar questions are found in transmission line analysis and geophysics. Mathematically, the presented scattering problems are described by the same hyperbolic system of partial differential equations.

The inverse scattering problem of determining the rod's impedance is of little practical interest as the impedance is either known or determined by other means. However, practical applications emerge if the system is extended by an impacting rod. For example, when using percussion drilling to generate blast holes for mining, an actuated piston transfers its kinetic energy to the rod during the impact. The induced force waves propagate along the rod until they reach the ground. The efficiency of the drilling process largely depends on the shape of the incident force waves at the ground. This motivates the development of a method that allows to prescribe the force over time at any axial rod position and to determine the corresponding impedance function of the impacting rod. A new method is necessary since the classical approaches of the scattering problem are not directly applicable to the impacting rod problem.

1.1 State of Research

In Fig. 1.1, an overview of the state of research is given. The classical 1D scattering problem consists of applying an impulse $I(0,t)$ at the boundary of the system and measuring the impulse response $O(0,t)$. As the input and output functions are known, the properties of the system may be reconstructed. The wave supporting system is described by two partial differential equations with coefficients $\gamma(x)$ and $\zeta(x)$ [46]. Physically, the coefficients refer to the distributed parameters of the system. Depending on the interpretation of the coefficients, the model is applicable in many fields of engineering. Some examples are the determination of the earth properties in geophysics [8, 10, 63, 64], the determination of the vocal tract shape by acoustical measurements at the lips [4, 65, 68 – 70] and the analysis of non-uniform transmission lines [12, 15, 51, 52].

There are different names for the partial differential equations in literature [11] depending on the parametrization. The *telegrapher's equations* result from the original equations by carrying out a coordinate transformation and introducing the impedance. After normalizing the *telegrapher's equations*, the equations in the form of *Schrödinger equations* arise by introducing two potentials. For expressing the variables $I(x,t)$ and $O(x,t)$ in terms of left and right propagating waves, no established term exists in literature. However, it is sometimes called *two component wave system*. Since the equations may be transformed into the original system, they all rely on the same mathematical foundation.

As many engineering fields are involved in finding solutions to the inverse scattering problem, the list of published literature is large. A comprehensive overview of many scattering problems is presented in [30] and [11]. The idea of solving the scattering problem using perturbation theory [54] has not been developed further since it has several disadvantages. The most serious ones are that the method does not make use of the structure of the problem and that the boundary conditions have to be known. The most promising methods are either assigned to differential methods or integral

1D Scattering Problem Description

Wave supporting system

$I(0,t)$

$O(0,t)$

$$\frac{\partial I(x,t)}{\partial t} = -\gamma(x)\frac{\partial O(x,t)}{\partial x}$$

$$\frac{\partial O(x,t)}{\partial t} = -\zeta(x)\frac{\partial I(x,t)}{\partial x}$$

x

Different Parametrization in Literature

- Telegrapher's equations (parametrized by local impedance)
- Schrödinger equations (parametrized by two potentials)
- Two component wave system (expressed in left and right traveling waves)

Solution Methods for Scattering Problems

Differential Methods

Continuous

- Layer peeling method

Discrete

- Dynamic deconvolution method
- Downward continuation method

Integral Equations

Examples

- Gelfand-Levitan
- Krein
- Gopinath-Sondhi

Extension to Impacting Rods (Impactor Synthesis Problem)

Based on Differential Methods

Thesis content:
Development of a method

Based on Integral Equations

Lundberg and Lesser

Experimental Validation

Thesis content:
Validation of 1D theory for impacting rods with jumps in cross section

Figure 1.1: Overview of the state of research and the content of the thesis.

equations. Historically, the first complete solution of the scattering problem has been documented by GELFAND and LEVITAN in terms of an integral equation. Subsequently, many similar integral equations have been published, e.g. in references [3, 68, 74]. They mainly differ in the input/output pair that is used in the formulation. In general, the computational costs of the integral equations is $O(\bar{n}^3)$, with $\bar{n}$ being the number of discretized time intervals. However, by making use of the special structure of the problem, the costs are reduced to $O(\bar{n}^2)$ [5].

The differential methods directly exploit the knowledge of the wave propagation process. As the material properties of the medium only vary in one direction, the medium is interpreted as layers of constant material properties. The further the layers are away from the force excitation the longer it takes to reach them. This also applies to the measured signal which is influenced by the layers at different moments for the first time. These observations are utilized in differential methods which determine the physical properties of the medium layer for layer [14, 62, 69, 71]. For a continuous change of the layers, the method is usually called *layer peeling method* [11]. The method with discrete changes of the layers is known as the *dynamic deconvolution method* [60] or as the *downward continuation method* [14].

The presented methods are also applicable to the longitudinal wave propagation in elastic rods. The classic scattering problem of determining the impedance of the rod by its impulse response is of little practical interest because the impedance is either known or determined by other means. However, the methods for the scattering problem may be extended to impacting rods where the impedance of one rod is determined so that a desired impact force is generated (*impactor synthesis problem*). This problem is motivated by theoretical work of LUNDBERG and COLLET who have shown that the efficiency of the percussion drilling process is increased if the incident wave is of exponential shape [49]. LUNDBERG and LESSER have extended the integral equations of GOPINATH and SONDHI [68] to solve the *impactor synthesis problem* in terms of an integral equation [47]. DUTTA has suggested to calculate the rod's impedance by systematically solving the direct problem and determining the solution which fits best to the desired impact force [24].

The *impactor synthesis problem* also occurs in dynamic material testing using a Split Hopkinson Pressure Bar apparatus. The test rig was invented by KOLSKY [40] and has been applied to determine the material properties of a test specimen at high strain rates [73]. The setup consists of three bars (striker, incident, transmitter) and a test specimen which is located between the incident and the transmitter bar. After the striker hits the incident bar, stress waves propagate along the incident bar toward the test specimen. At the test specimen, the stress waves are partly reflected and transmitted. The incident, reflected, and transmitted waves are recorded by strain gauges that are attached to the incident and transmitter bar. For the evaluation of the detected waves, several methods have been proposed, e.g. in references [29, 33, 55, 57].

It might be attempted to generate an incident wave that produces a constant strain rate in the specimen. An evident difficulty is that this wave depends on the unknown properties of the specimen. One possibility to fulfill this requirement is to optimize the geometry of the striker [22, 26, 45, 75] accordingly. Another approach is to introduce a shaping bar between striker and incident bar [28]. So far, the optimized striker geometries are designed by experience and knowledge of the solution of the direct problem. However, the determination of the geometry by solving the corresponding inverse problem promises to deliver faster and more accurate results.

Apart from the two presented approaches, no other methods are known in literature which solve the *impactor synthesis problem*. Specifically, no methods are known which are based on the idea of differential methods to recursively determine the rod's impedance layer by layer.

The quality of the approaches strongly relies on the validity of the 1D model. Pochhammer [58] and Chree [21] independently formulated the exact solution of the three-dimensional wave propagation problem in an infinitely long cylinder. The solution relates the phase velocity of the longitudinal waves to the wavelength of its propagating modes (dispersion relation) [6]. The dispersive effect increases the greater the ratio between rod diameter and wavelength is [27]. To correct these effects, several methods have been established, e.g. in references [25, 53].

Since the dispersive effects are not represented by the 1D model, it has to be examined under which circumstances the 1D model delivers accurate results. It has been proved that the 1D results yield good accordance with their corresponding experimental results, e.g. in reference [56]. However, only standard geometries (constant, conical, etc.) have been studied. It has not been systematically examined yet if a complex variation of the rod geometry, including jumps in cross section area, delivers good accordance with the experimental results.

1.2 Thesis Purpose and Structure

The literature research reveals that there are very few methods which cover the *impactor synthesis problem*. One approach is to systematically optimize the rod's impedance by solving the direct problem. This approach has several drawbacks: On the one hand, the method does not guarantee that the solution converges to the exact solution and on the other hand, the computational costs are very high. A more promising approach has been developed by LUNDBERG and LESSER [47] based on the integral equations of GOPINATH and SONDHI [68]. This method provides a solution in two steps: Initially, a deconvolution is carried out to determine the impulse response of the unknown rod. Subsequently, the impulse response is substituted into an integral equation which is finally evaluated for the determination of the unknown rod impedance. However, several approximation methods are applied to obtain the rod's impedance. Firstly, the impulse response of the given rod, which is part of the convolution equation, is not always known analytically and has to be approximated. Secondly, approximation methods are applied to deconvolve the convolution equation. And thirdly, the integral equation is solved approximately.

As compared to the evaluation of integral equations, a differential-based method, that exploits the structure of the problem, merely has to approximate the unknown rod's impedance by piecewise constant layers. However, the existing methods to solve the scattering problem are not expandable to the *impactor synthesis problem* straightforwardly.

The **main purpose** of the thesis is to develop a method that solves the *impactor synthesis problem* based on the ideas of the differential methods of the scattering problem. A **further aim** is to validate the underlying 1D model experimentally for an impact with complex rod geometries.

In the thesis, the setups presented in Fig. 1.2 are investigated. In Setup 1, a rod with constant cross section area is impacting a rod with exponentially increasing cross section area whose right end is fixed. Setup 2 is a textbook example where a rod consisting of two cross section area sections is hitting a rod with a constant cross section area. Setup 3 is more complex, including variations in geometry and the consideration of external friction forces. Setup 4/5 only differ in the prescribed impact force N_{I} that shall be generated by the impacting rod whose impedances are determined accordingly. In Setup 6, a rod with two steps in cross section area is hitting a rod with a constant cross section area. Setup 7 is similar to Setup 6 but with more complex cross section area

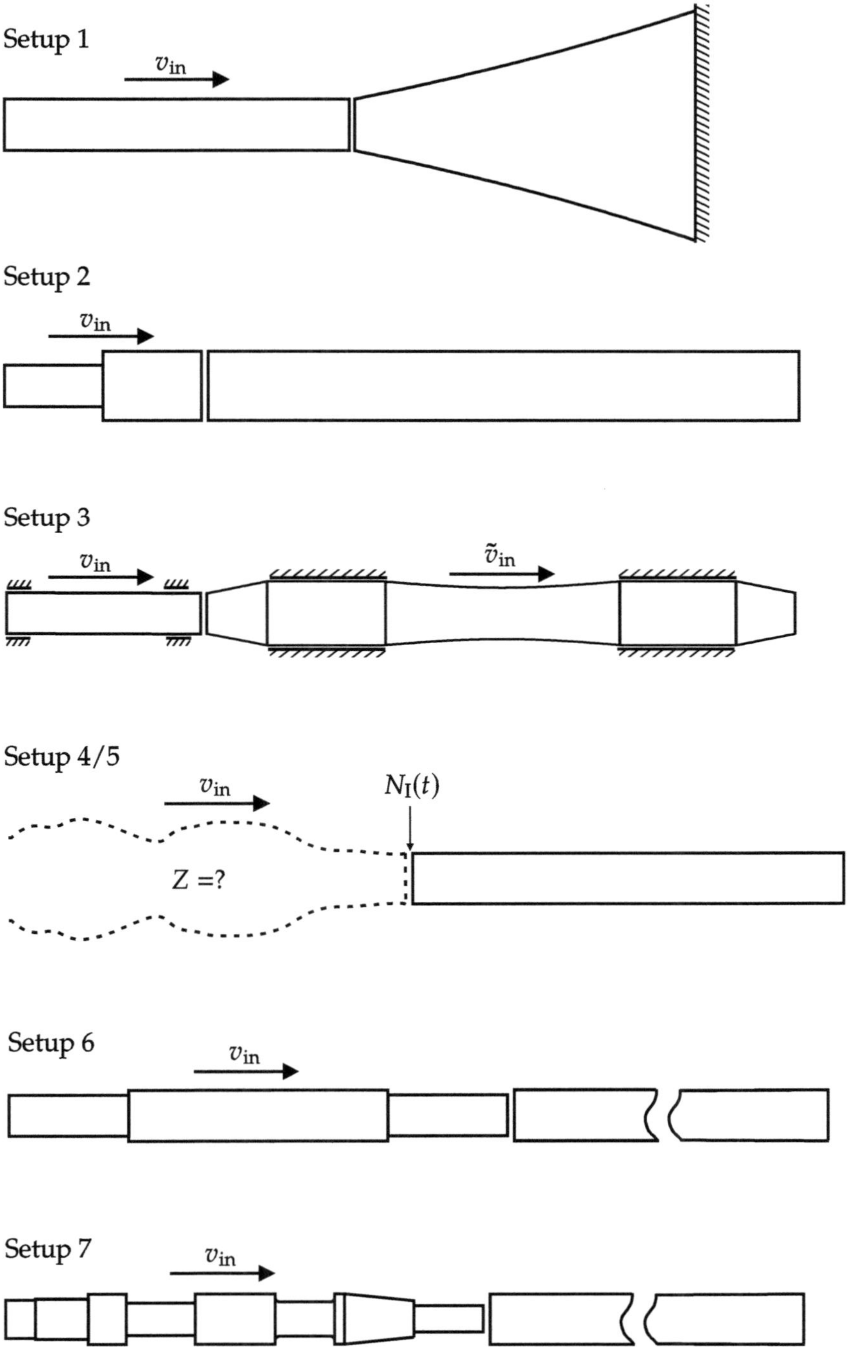

Figure 1.2: Investigated setups.

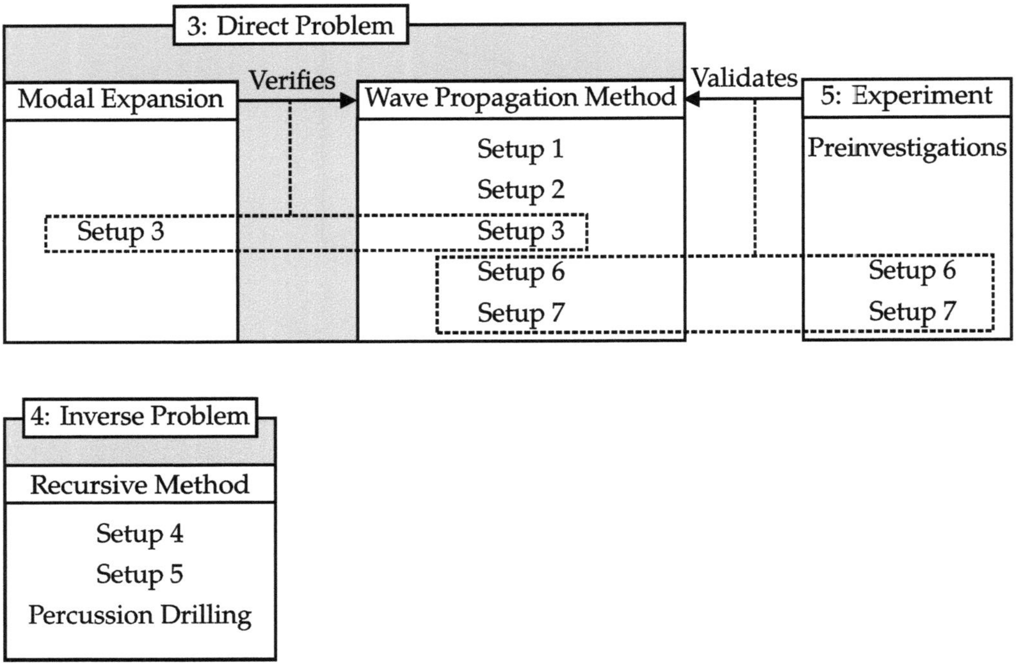

Figure 1.3: Overview of the content of the chapters.

variations of the impacting rod. The setups are investigated by different means. They are either investigated by applying the modal expansion method, the wave propagation method, the recursive method or experimentally (see Fig. 1.3). The main purpose of the thesis is to develop the recursive method which solves the *impactor synthesis problem*. The recursive method is based on the wave propagation method. The modal expansion method verifies the wave propagation method whereas the experimental results validate the underlying 1D model. This motivates the structure of the following chapters:

Chapter 2: In this chapter, the fundamentals that are used in the following chapters are presented.

Chapter 3: The wave propagation method for the calculation of 1D wave propagation problems in non-uniform rods is introduced. The range of applications is demonstrated by applying the method to Setup 1 and Setup 2. The method is verified by a comparison of results obtained with the modal expansion method versus results obtained with the wave propagation method applied to Setup 3.

Chapter 4: The fourth chapter covers the new recursive method. The presentation of the main idea is followed by its mathematical description. A condition is derived to assess the existence of a solution of the inverse problem for every step. Afterward, the working principle of the recursive method is presented by employing Setup 4/5. Setup 4 only differs from Setup 5 in the prescribed impact force N_I. Depending on the impact force, a solution of the inverse problem exists (Setup 4) or not (Setup 5). Finally, the method is applied in percussion drilling.

Chapter 5: The experimental setup is presented. Several experiments are conducted using Setup 6 to determine the influence factors on the measurements and to examine the repeatability of the results. Subsequently, the simulation and experimental results obtained with Setup 6 and Setup 7 are analyzed and compared. Thus, the underlying 1D model is validated.

Chapter 6: In the last chapter, the thesis is concluded.

2 Fundamentals

In this chapter, the fundamentals, which are applied in the subsequent chapters, are presented. Initially, the governing partial differential equation (PDE) for the 1D longitudinal wave propagation in rods is derived. It is shown that the PDE can be solved analytically for a special class of cross sectional area variations by applying the modal expansion method. However, the respective results are not expressable in closed-form but with an infinite series of eigenfunctions and eigenfrequencies. The modal expansion method reveals two drawbacks: Firstly, the analytical solution applies only to very few cross sectional area variations. And secondly, many series terms have to be considered to approach the exact solution. The solutions obtained by the modal expansion method are used to verify the results of the wave propagation method whose main idea is to subdivide the impacting rods into a large number of elements of piecewise constant impedance. For these elements, the wave equation can be solved easily. Its D'ALEMBERT solution is discussed in detail. Especially, the transition and boundary conditions are considered as they are essential to solve the inverse problem.

2.1 Derivation of the Governing PDE

In Fig. 2.1a, a schematic sketch of a rod is depicted. The mass density ρ, the Young's modulus E and the cross section area A only vary in x-direction. The axial displacement of the particles at position x is described by $u(x,t)$. It is assumed that the centroid of the cross section area lies on a straight line perpendicular to the cross section area [36]. Applying NEWTON's second law on the free body diagram of a rod segment (see Fig. 2.1b) leads to

$$\frac{\partial N(x,t)}{\partial x} = \rho(x)A(x)\frac{\partial^2 u(x,t)}{\partial t^2}, \tag{2.1}$$

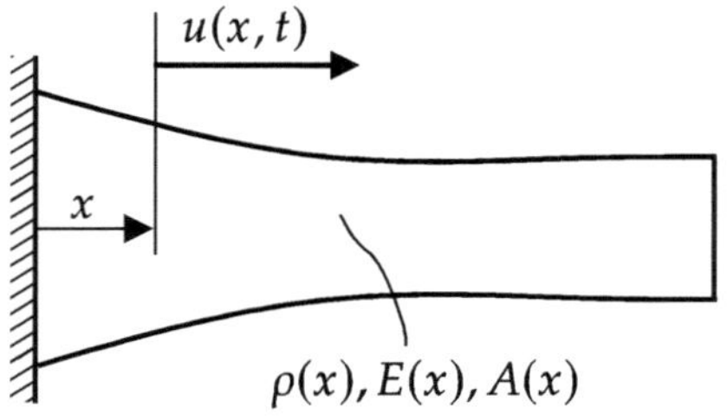

(a) Schematic sketch of a rod.

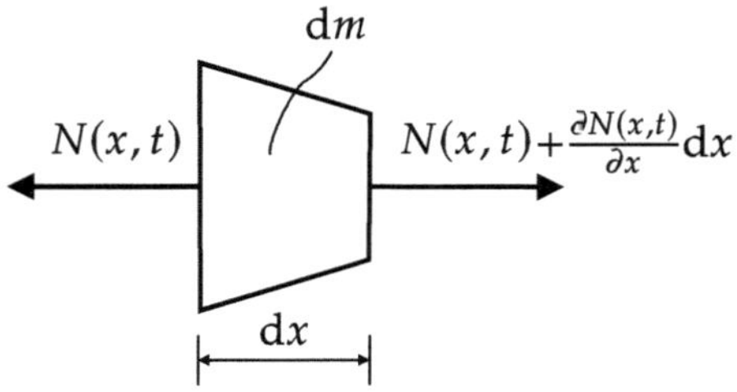

(b) Free body diagram of a rod segment.

Figure 2.1: (a) Schematic sketch of a rod and (b) the free body diagram of a rod segment for the derivation of the governing PDE.

where $N(x,t)$ is the normal force, positive in tension [35]. The particle velocity

$$v(x,t) = \frac{\partial u(x,t)}{\partial t} \tag{2.2}$$

is obtained by differentiating the displacement with respect to time whereas the axial strain

$$\varepsilon(x,t) = \frac{\partial u(x,t)}{\partial x} \tag{2.3}$$

is obtained by differentiating the displacement with respect to space [50]. Inserting the force relation

$$N(x,t) = A(x)\sigma(x,t) \tag{2.4}$$

together with the relation between axial stress $\sigma(x,t)$ and strain $\varepsilon(x,t)$ (Hooke's law)

$$\sigma(x,t) = E(x)\varepsilon(x,t) = E(x)\frac{\partial u(x,t)}{\partial x} \tag{2.5}$$

into Eq. (2.1) yields the governing PDE

$$\frac{\partial}{\partial x}\left[E(x)A(x)\frac{\partial u(x,t)}{\partial x}\right] = \rho(x)A(x)\frac{\partial^2 u(x,t)}{\partial t^2} \tag{2.6}$$

for the longitudinal wave propagation in rods. The model is equivalently describable by a hyperbolic system of linear partial differential equations

$$\begin{aligned}\frac{\partial N(x,t)}{\partial t} &= E(x)A(x)\frac{\partial v(x,t)}{\partial x},\\ \frac{\partial v(x,t)}{\partial t} &= \frac{1}{\rho(x)A(x)}\frac{\partial N(x,t)}{\partial x}.\end{aligned} \tag{2.7}$$

At this point, it is advantageous to introduce the impedance [32]

$$Z(x) = A(x)\rho(x)c(x) \tag{2.8}$$

with

$$c(x) = \sqrt{\frac{E(x)}{\rho(x)}} \tag{2.9}$$

being the wave propagation speed. The impedance of a rod is defined as the ratio between the force amplitude and its related velocity amplitude [36]. Thus, the impedance may be interpreted as the resistance of the elastic structure to the acting forces.

Introducing the coordinate

$$\theta(x) = \int_0^x \frac{1}{c(x')}\,\mathrm{d}x' \tag{2.10}$$

together with the notation $v(x(\theta),t) = \bar{v}(\theta,t)$ etc. enables to transform the system (2.7) into [46]

$$\begin{aligned}\frac{\partial \bar{N}(\theta,t)}{\partial t} &= \bar{Z}(\theta)\frac{\partial \bar{v}(\theta,t)}{\partial \theta},\\ \frac{\partial \bar{v}(\theta,t)}{\partial t} &= \frac{1}{\bar{Z}(\theta)}\frac{\partial \bar{N}(\theta,t)}{\partial \theta}.\end{aligned} \tag{2.11}$$

Therefore, the equations are describable solely in dependance of the characteristic impedance.

2.2 Analytical Solution of the Governing PDE

In general, it is not possible to solve the governing PDE analytically. For a special class of cross sectional area variations and for homogeneous materials analytical solutions exist [1, 7, 43, 44, 59]. Guo and Yang [34] have shown that the cross sectional area variation has to be either square, exponential, square of hyperbolic cosine or square of cosine. The corresponding approach is named as the modal expansion method. Its main ideas are presented below.

For homogeneous materials, the governing PDE (Eq. (2.6)) is rewritten

$$\frac{\partial^2 u(x,t)}{\partial x^2} + h(x)\frac{\partial u(x,t)}{\partial x} - \frac{1}{c_0^2}\frac{\partial^2 u(x,t)}{\partial t^2} = 0, \;\; h(x) = \frac{\mathrm{d}A(x)/\mathrm{d}x}{A(x)}, \;\; c_0^2 = \frac{E}{\rho}. \tag{2.12}$$

Inserting the exponential approach

$$u(x,t) = a(x)\,\mathrm{e}^{i(kx-\omega t)} \tag{2.13}$$

with wave amplitude $a(x)$, wave number k and angular frequency ω in Eq. (2.12) leads to two equations

$$\text{Im: } 2\frac{\mathrm{d}a(x)}{\mathrm{d}x} + h(x)a(x) = 0, \tag{2.14}$$

$$\text{Re: } \frac{\mathrm{d}^2 a(x)}{\mathrm{d}x^2} - k^2 a(x) + h(x)\frac{\mathrm{d}a(x)}{\mathrm{d}x} + \frac{\omega^2}{c_0^2}a(x) = 0 \tag{2.15}$$

for the imaginary (Im) and real (Re) part, respectively. From the first equation, $a(x)$ is determined

$$a(x) = a_0\sqrt{\frac{A_0}{A(x)}}, \tag{2.16}$$

where a_0 and A_0 are the wave amplitude and cross section area at $x = 0$. By eliminating $a(x)$ in Eq. (2.15), the equation

$$k^2 = \frac{\omega^2}{c_0^2} - \underbrace{\left(\frac{1}{4}h(x)^2 + \frac{1}{2}\frac{\mathrm{d}h(x)}{\mathrm{d}x}\right)}_{b} \tag{2.17}$$

for k is obtained. Analytical solutions are found if the wave number is independent of x which means that b has to be constant. This applies only to very few cross section areas, including the ones mentioned above. For those cross section areas the phase velocity

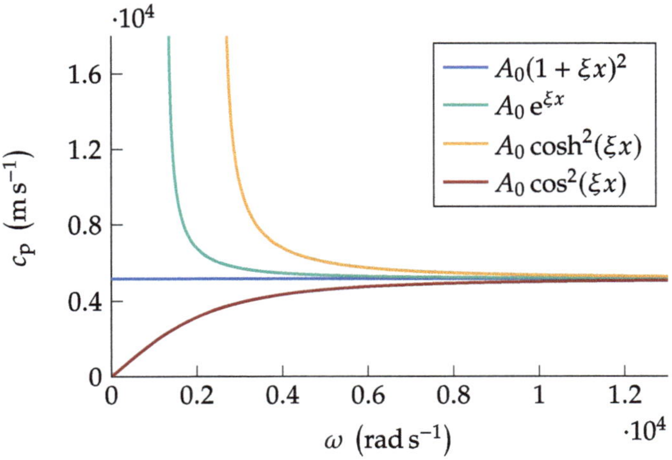

Figure 2.2: Phase speed for $\xi = 1$ of the cross section areas that meet the requirement for an analytical solution (constant k).

$$c_p = \frac{\omega}{k} = \frac{c_0}{\sqrt{1 - \frac{c_0^2}{\omega^2} b}} \tag{2.18}$$

depends generally on the angular frequency ω which can be seen in Fig. 2.2. Except from the curve related to quadratic cross sectional area variation, all the other curves are dispersive. Moreover, the exponential and square of hyperbolic cosine cross sectional area variations exhibit a cutoff frequency ω_c that is calculated with

$$\omega_c = c_0 \sqrt{b}. \tag{2.19}$$

Therefore, these cross section areas are mechanical high pass filters. In Tab. 2.1, the most important values related to the analytical solutions are summarized. These values are the cross section area variation $A(x)$, the constant b, the cutoff frequency ω_c, the phase velocity c_p and the wave amplitude $a(x)$.

$A(x)$	b	ω_c	k	c_p	$a(x)$
$A_0(1+\xi x)^2$	0	0	$\frac{\omega}{c_0}$	c_0	$\frac{a_0}{1+\xi x}$
$A_0\,\mathrm{e}^{\xi x}$	$\frac{\xi^2}{4}$	$c_0\frac{\xi}{2}$	$\sqrt{\frac{\omega^2}{c_0^2}-\frac{\xi^2}{4}}$	$\frac{c_0}{\sqrt{1-\left(\frac{c_0\xi}{2\omega}\right)^2}}$	$a_0\,\mathrm{e}^{-\frac{\xi}{2}x}$
$A_0\cosh^2(\xi x)$	ξ^2	$c_0\xi$	$\sqrt{\frac{\omega^2}{c_0^2}-\xi^2}$	$\frac{c_0}{\sqrt{1-\left(\frac{c_0\xi}{\omega}\right)^2}}$	$\frac{a_0}{\cosh(\xi x)}$
$A_0\cos^2(\xi x)$	$-\xi^2$	0	$\sqrt{\frac{\omega^2}{c_0^2}+\xi^2}$	$\frac{c_0}{\sqrt{1+\left(\frac{c_0\xi}{\omega}\right)^2}}$	$\frac{a_0}{\cos(\xi x)}$

Table 2.1: Properties of the analytical solutions, based on [34].

2.3 Wave Equation

The governing PDE (Eq. (2.6)) simplifies for homogeneous materials and constant cross section areas to the well known wave equation [23, 41]

$$\frac{\partial^2 u(x,t)}{\partial t^2} = c^2\frac{\partial^2 u(x,t)}{\partial x^2},\quad c^2 = \frac{E}{\rho}. \tag{2.20}$$

The wave equation also applies to all physical values that are proportional to a partial derivative of u, especially, the velocity

$$v(x,t) = \frac{\partial u(x,t)}{\partial t}, \tag{2.21}$$

the strain

$$\varepsilon(x,t) = \frac{\partial u(x,t)}{\partial x}, \tag{2.22}$$

and the normal force

$$N(x,t) = EA\frac{\partial u(x,t)}{\partial x} \tag{2.23}$$

as can be proven by partially differentiating the wave equation with respect to t and x [72].

2.3.1 D'Alembert Solution

Initially, the wave propagation along an infinitely long rod without any transition or boundary conditions is considered. The two arbitrary functions $f(x-ct)$ and $g(x+ct)$ with argument $x-ct$ and $x+ct$ solve the wave equation [2]. Due to the linearity of the PDE, the sum of the functions also solves the wave equation and the general solution of the displacement is given by

$$u(x,t) = f(x-ct) + g(x+ct). \tag{2.24}$$

The functions f and g describe waves traveling at wave propagation speed c in positive and negative x–direction, respectively.

For a unique solution of the functions f and g, the initial conditions have to be evaluated. Let

$$u_0(x) = u(x,0), \;\; v_0(x) = \left.\frac{\partial u(x,t)}{\partial t}\right|_{t=0} \tag{2.25}$$

be the initial displacement and velocity for $-\infty \leq x \leq \infty$ at $t = 0$. Then, the initial value problem is solved with [72]

$$u(x,t) = \frac{1}{2}\left[u_0(x-ct) + u_0(x+ct) + \frac{1}{c}\int_{x-ct}^{x+ct} v_0(\bar{x})\,\mathrm{d}\bar{x}\right]. \tag{2.26}$$

The initial displacement $u_0(x)$ splits up in two parts of half magnitude each of which are traveling in positive and negative x–direction. Moreover, due to the finite wave propagation speed, the initial velocity at $x = \bar{x}$ only influences the displacements in the interval $x \in [\bar{x} - ct, \bar{x} + ct]$.

2.3.2 Transition and Boundary Conditions

For practical applications, transition and boundary conditions have to be considered. An overview of the most common transition and boundary conditions is given in Fig. 2.3.

Transition Conditions

The aim is to find a coherence between the known incident force wave $N_\mathrm{i}(x_\mathrm{tr},t)$ and its reflected $N_\mathrm{r}(x_\mathrm{tr},t)$ and transmitted parts $N_\mathrm{t}(x_\mathrm{tr},t)$. Therefore, an incident wave

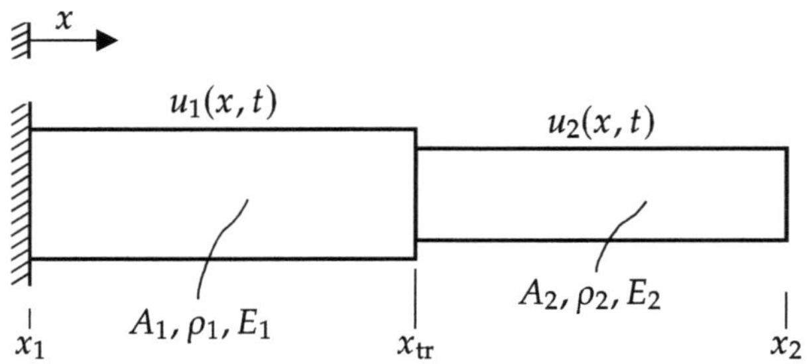

Figure 2.3: Two homogeneous elements with a fixed end at $x = x_1$, a free end at $x = x_2$ and the transition at $x = x_{\mathrm{tr}}$.

$f_{\mathrm{i}}(x - c_1 t)$, which travels in the first element toward the transition, is considered. At the transition, the incident wave is partly reflected and transmitted. This leads to traveling waves $g_{\mathrm{r}}(x + c_1 t)$ and $f_{\mathrm{t}}(x - c_2 t)$ in the first respectively second element. In general, the displacements of the two elements are described by

$$u_k(x,t) = f_k(x - c_k t) + g_k(x + c_k t), \;\; k = 1,2, \tag{2.27}$$

which simplifies for the given consideration to

$$u_1(x,t) = f_{\mathrm{i}}(x - c_1 t) + g_{\mathrm{r}}(x + c_1 t), \tag{2.28}$$
$$u_2(x,t) = f_{\mathrm{t}}(x - c_2 t). \tag{2.29}$$

Moreover, two transition conditions have to be stated which are continuity of displacement

$$u_1\left(x_{\mathrm{tr}}, t\right) = u_2\left(x_{\mathrm{tr}}, t\right) \tag{2.30}$$

and force equilibrium

$$N_{\mathrm{i}}(x_{\mathrm{tr}}, t) + N_{\mathrm{r}}(x_{\mathrm{tr}}, t) = N_{\mathrm{t}}(x_{\mathrm{tr}}, t). \tag{2.31}$$

The displacement $u_1(x_{\mathrm{tr}}, t)$ is composed of an incident part $u_{\mathrm{i}}(x_{\mathrm{tr}}, t)$ and a reflected one $u_{\mathrm{r}}(x_{\mathrm{tr}}, t)$, whereas $u_2(x_{\mathrm{tr}}, t)$ only consists of the transmitted part $u_{\mathrm{t}}(x_{\mathrm{tr}}, t)$. In terms of traveling waves, Eq. (2.30) is written as

$$\underbrace{f_{\mathrm{i}}\left(x_{\mathrm{tr}} - c_1 t\right)}_{u_{\mathrm{i}}(x_{\mathrm{tr}},t)} + \underbrace{g_{\mathrm{r}}\left(x_{\mathrm{tr}} + c_1 t\right)}_{u_{\mathrm{r}}(x_{\mathrm{tr}},t)} = \underbrace{f_{\mathrm{t}}\left(x_{\mathrm{tr}} - c_2 t\right)}_{u_{\mathrm{t}}(x_{\mathrm{tr}},t)}. \tag{2.32}$$

As the displacement at $x = x_{tr}$ is continuous, this is also true for the velocity

$$\left.\frac{\partial u_1(x,t)}{\partial t}\right|_{x=x_{tr}} = \left.\frac{\partial u_2(x,t)}{\partial t}\right|_{x=x_{tr}}, \tag{2.33}$$

$$-c_1 f_i'(x_{tr} - c_1 t) + c_1 g_r'(x_{tr} + c_1 t) = -c_2 f_t'(x - c_2 t), \tag{2.34}$$

$$-c_1 \underbrace{E_1 A_1 f_i'(x_{tr} - c_1 t)}_{N_i(x_{tr},t)} + c_1 \underbrace{E_1 A_1 g_r'(x_{tr} + c_1 t)}_{N_r(x_{tr},t)} = -c_2 \frac{E_1 A_1}{E_2 A_2} \underbrace{E_2 A_2 f_t'(x_{tr} - c_2 t)}_{N_t(x_{tr},t)} \tag{2.35}$$

from which

$$-N_i(x_{tr}, t) + N_r(x_{tr}, t) = -\frac{Z_1}{Z_2} N_t(x_{tr}, t) \tag{2.36}$$

follows. Equation (2.36), together with Eq. (2.31) results in the final expressions

$$N_t(x_{tr}, t) = \frac{2Z_2}{Z_2 + Z_1} N_i(x_{tr}, t) = T_{1,2} N_i(x_{tr}, t), \tag{2.37}$$

$$N_r(x_{tr}, t) = \frac{Z_2 - Z_1}{Z_2 + Z_1} N_i(x_{tr}, t) = R_{1,2} N_i(x_{tr}, t) \tag{2.38}$$

for the transmitted and reflected force waves depending on the incident force. The transmission and reflection factors from element i to element j are defined by

$$T_{i,j} = \frac{2Z_j}{Z_j + Z_i}, \tag{2.39}$$

$$R_{i,j} = \frac{Z_j - Z_i}{Z_j + Z_i}. \tag{2.40}$$

Boundary Conditions

Along with the transition conditions, the boundary conditions are necessary to determine the solution for $u(x, t)$. In Fig. 2.3, the boundary conditions are a fixed end at $x = x_1$ and a free end at $x = x_2$.

At the fixed end, the displacement is identical zero, i. e.

$$u(x_1, t) \equiv 0, \tag{2.41}$$

whereas at the free end, the force is identical zero, i. e.

$$N(x_2, t) \equiv 0. \tag{2.42}$$

The corresponding reflected forces at the fixed respectively free end are obtained by taking the limit $Z_2 \rightarrow \infty$ (fixed end) or setting $Z_2 = 0$ (free end) in Eq. (2.38)

$$N_\mathrm{r}(x_1, t) = N_\mathrm{i}(x_1, t), \tag{2.43}$$

$$N_\mathrm{r}(x_2, t) = -N_\mathrm{i}(x_2, t). \tag{2.44}$$

3 Impacting Rods: Direct Problem

The aim of this chapter is to verify the wave propagation method. Initially, the wave propagation method is presented. Its main equations for one time step are derived. Possible applications of the wave propagation method are demonstrated by two examples. Afterward, the wave propagation method is summarized. Subsequently, the modal expansion method is applied to a hit of two rods of constant and exponential increasing cross section area. After determining the eigenfunctions of the problem, the initial conditions are incorporated by using an orthogonality relation. At the end of the chapter, the results that are determined with both methods, are compared. It is shown that the wave propagation method provides more accurate results if a finite number of terms is used.

3.1 Wave Propagation Method

The direct solution of the impacting rod problem has been discussed extensively in literature, e.g. in references [31, 38]. One approach is to model the rods with concentrated masses and stiffnesses. This also applies to the well known finite element method (FEM). The FEM delivers accurate results in a wide range of applications. However, the method reveals disadvantages when wave problems are considered: On the one hand, the accelerations are assumed to be finite so that they are numerically integrable. This is the reason why strong discontinuities in force and velocity, which often occur in impacting problems, are not displayed correctly. On the other hand, the FEM does not take into account that the wavefront is traveling through the elements with finite speed. Thus, displacements are calculated at axial positions which are not reached by the wavefront yet. These disadvantages vanish when the system is modeled by distributed masses and stiffnesses which are described by the wave equation. This approach has been studied by Shorr in several publications, e.g. in references [66, 67]. The following derivation of the wave propagation method is partly based on his publications.

The main idea in modeling is to approximate the impedances of the rods by piecewise constant step functions. In Fig. 3.1, the elements of the rods are depicted for the special case of constant material parameters ρ and E. Therefore, the impedances only change due to the changing cross section area. This discretization is advantageous since analytical solutions exist on elements with constant impedance. Moreover, the wave propagation speed c_j is constant on each element j. The element lengths ℓ_j are adjusted so that the waves travel through every element during a constant time step Δt

$$\ell_j = \sqrt{\frac{E_j}{\rho_j}}\Delta t = c_j \Delta t. \tag{3.1}$$

Thus, all the traveling waves reach the boundaries of the corresponding elements at the same time.

If external forces or loads are applied, the following assumptions pertain:

- Line loads are modeled as concentrated external forces.
- External forces only act on the nodes between the elements.
- External forces may change in time but are constant during one time step.

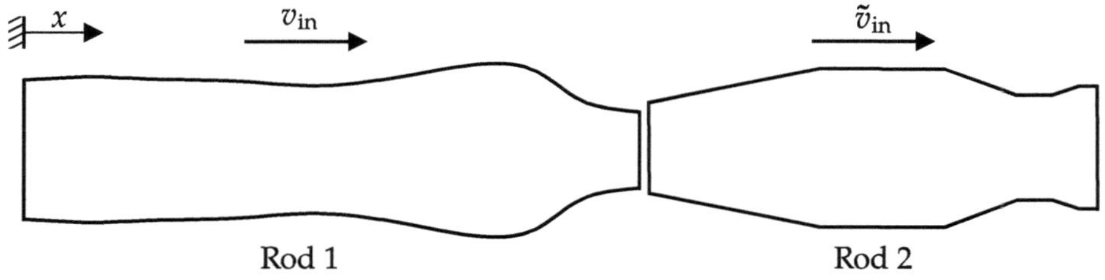

(a) Impact of non-uniform rods at initial velocities v_{in} in Rod 1 and $\tilde{v}_{\text{in}}$ in Rod 2.

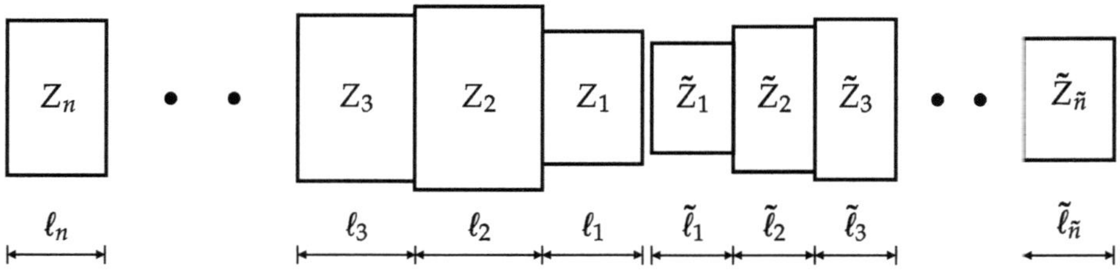

(b) Discretization of the rods with constant impedances.

Figure 3.1: Impact of two non-uniform rods and their modeling.

As long as the impact force is compressive, the rods are calculated as if they were stuck together. When the force tends to become tensile, the rods separate. Therefore, the rods are treated as two individual rods with free ends at their former hitting surfaces.

3.1.1 Equations for One Time Step

During one time step, the waves are traveling from one element border to the next element border. At the end of each time step at $t = t_0$, all elements are in equilibrium (see Fig. 3.2) which means that both the normal force N and the particle velocity v is constant within the elements. The objective of this section is to derive the equations which are required to determine the new equilibrium state one time step later at $t = t_0 + \Delta t$. To this end, the transition forces and velocities are determined first. This enables the calculation of the incident force and velocity waves starting from the element borders. The new forces and velocities are finally obtained by a superposition of the incident waves and their old values.

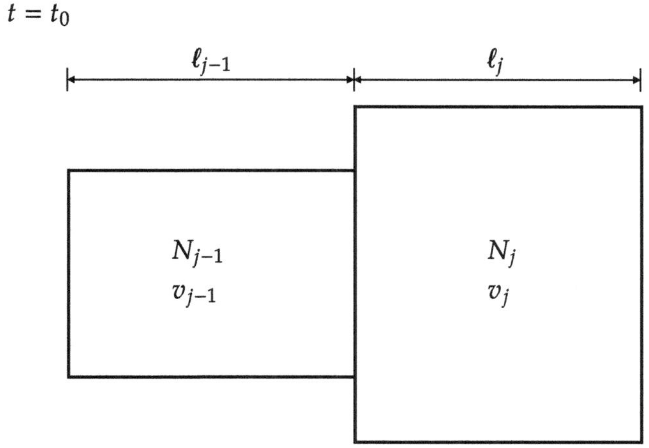

Figure 3.2: Two elements in equilibrium state at the end of each time step.

In general, the forces and velocities of two adjacent elements are not equal at $t = t_0$, that is $N_{j-1} \neq N_j$ and $v_{j-1} \neq v_j$. These inequalities cause waves to propagate into the elements starting from the transition (see Fig. 3.3a). The superscripts indicate that N or v is either acting on the right end (superscript +) or left end (superscript −) of the corresponding element.

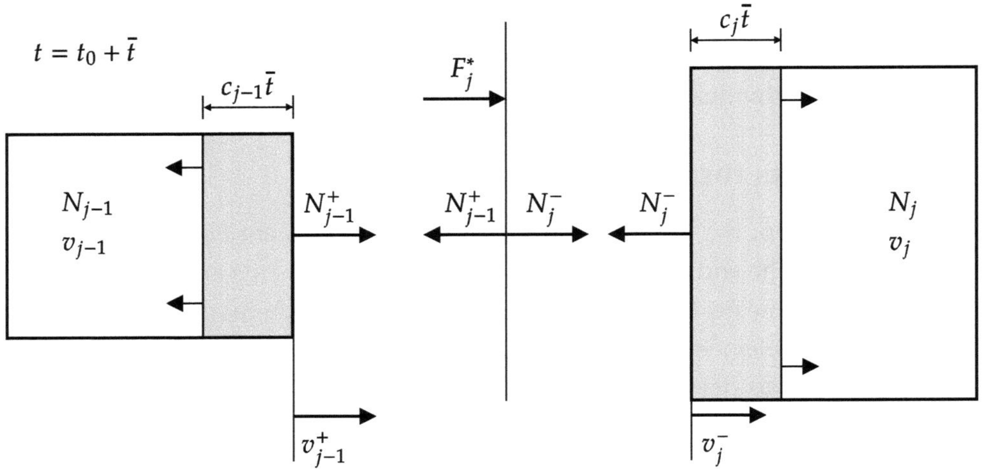

(a) Free body diagram at the transition of two elements during one time step.

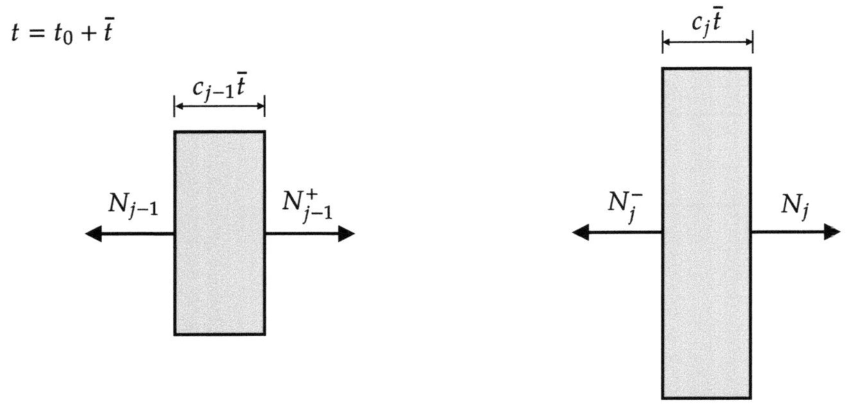

(b) Free body diagrams of the part that is covered by the incident wave of element $j-1$ and element j.

Figure 3.3: Free body diagrams at the transition.

At the element transitions, force equilibrium and continuity is claimed. As a consequence of continuity, the velocities of the related element particles have to be equal as well. Mathematically, the claims are expressed by

$$N_{j-1}^+ - N_j^- = F_j^*, \tag{3.2}$$

$$v_{j-1}^+ = v_j^- \tag{3.3}$$

with the external force F^* and the transition forces N^+_{j-1}, N^-_j and velocities v^+_{j-1}, v^-_j acting on the element borders. During the travel time $\bar{t} \in [t_0, t_0 + \Delta t]$, the waves have propagated the distances $c_{j-1}\bar{t}$ and $c_j\bar{t}$, respectively. In Fig. 3.3b, the free body diagrams of the parts that are covered by the incident waves are depicted. Applying the momentum conservation law on the free body diagram of element j

$$\begin{aligned} \left(N_j - N^-_j\right)\bar{t} &= A_j\rho_j c_j\bar{t}\left(v^-_j - v_j\right), \\ N^-_j &= N_j - Z_j\left(v^-_j - v_j\right) \end{aligned} \tag{3.4}$$

leads to an equation for the transition force acting on the left end of element j. Analogously, the transition force acting on the right end of element $j-1$ is derived

$$N^+_{j-1} = N_{j-1} + Z_{j-1}\left(v^+_{j-1} - v_{j-1}\right). \tag{3.5}$$

By evaluating the transition conditions (Eqs. (3.2), (3.3)) together with the above derived equations (Eqs. (3.4), (3.5)), the unknown transition forces N^+_j, N^-_j and velocities v^+_j, v^-_j

$$N^+_j = \frac{\left(\left(v_{j+1} - v_j\right)Z_{j+1} + N_{j+1} + F^*_{j+1}\right)Z_j + N_j Z_{j+1}}{Z_{j+1} + Z_j}, \tag{3.6}$$

$$v^+_j = \frac{Z_j v_j + Z_{j+1}v_{j+1} - N_j + N_{j+1} + F^*_{j+1}}{Z_{j+1} + Z_j}, \tag{3.7}$$

$$N^-_j = \frac{\left(\left(v_j - v_{j-1}\right)Z_{j-1} + N_{j-1} - F^*_j\right)Z_j + N_j Z_{j-1}}{Z_{j-1} + Z_j}, \tag{3.8}$$

$$v^-_j = \frac{Z_j v_j + Z_{j-1}v_{j-1} + N_j - N_{j-1} + F^*_j}{Z_{j-1} + Z_j} \tag{3.9}$$

are determined after the subscripts have been manipulated. The force difference ΔN between the transition forces N^+_j, N^-_j and the old force N_j leads to traveling force waves

$$\Delta N_{j,\mathrm{L}} = N^+_j - N_j, \tag{3.10}$$

$$\Delta N_{j,\mathrm{R}} = N^-_j - N_j \tag{3.11}$$

as can be seen in Fig. 3.4. The subscripts denote if the waves are traveling to the left

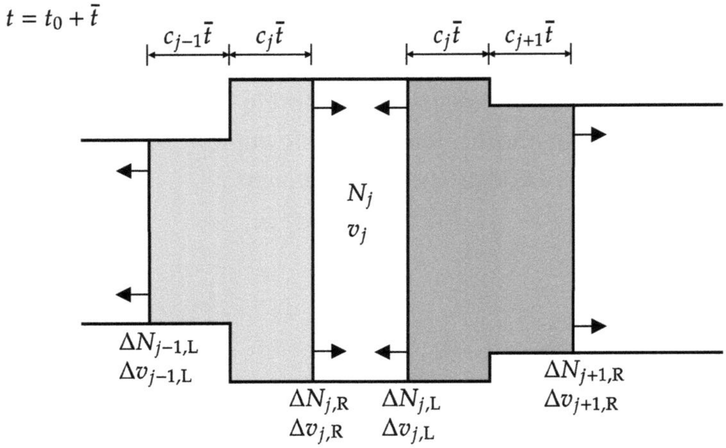

Figure 3.4: Incident waves in element j and in its adjacent elements.

(subscript L) or right (subscript R). Analogously, the difference in velocity Δv leads to traveling velocity waves

$$\Delta v_{j,\mathrm{L}} = v_j^+ - v_j, \tag{3.12}$$

$$\Delta v_{j,\mathrm{R}} = v_j^- - v_j. \tag{3.13}$$

The new force $N_{j,\mathrm{new}}$ and velocity $v_{j,\mathrm{new}}$ are finally obtained by a superposition of the incident waves and the old values

$$N_{j,\mathrm{new}} = N_j + \Delta N_{j,\mathrm{L}} + \Delta N_{j,\mathrm{R}}, \tag{3.14}$$

$$v_{j,\mathrm{new}} = v_j + \Delta v_{j,\mathrm{L}} + \Delta v_{j,\mathrm{R}}. \tag{3.15}$$

3.1.2 Boundary Conditions

The most common boundary conditions (given boundary force, given velocity/displacement, energy sink) are presented.

For **given boundary forces** $N_n^-, N_{\tilde{n}}^+$, the subsequent equations for the velocities are derived by manipulating Eqs. (3.4) and (3.5)

$$v_n^- = \frac{N_n - N_n^-}{Z_n} + v_n, \tag{3.16}$$

$$v_{\tilde{n}}^+ = \frac{N_{\tilde{n}}^+ - N_{\tilde{n}}}{Z_{\tilde{n}}} + v_{\tilde{n}}. \tag{3.17}$$

For **given velocities** $v_n^-, v_{\tilde{n}}^+$, the resulting forces are determined by rearranging Eqs. (3.16) and (3.17)

$$N_n^- = N_n + Z_n \left(v_n - v_n^-\right), \tag{3.18}$$

$$N_{\tilde{n}}^+ = N_{\tilde{n}} + Z_{\tilde{n}} \left(v_{\tilde{n}}^+ - v_{\tilde{n}}\right). \tag{3.19}$$

For a **given energy sink**, i.e. that no reflecting waves occur at the boundaries, the equations are

$$N_n^- = N_n, \qquad N_{\tilde{n}}^+ = N_{\tilde{n}}. \tag{3.20}$$

The corresponding velocities are obtained by substituting the determined forces into Eqs. (3.16) and (3.17).

3.1.3 Examples

The range of applications of the wave propagation method is demonstrated by applying the method to two examples. The first example is a textbook example with one jump in impedance. The second example is more complex, including variations in geometry and material parameters. Moreover, friction is considered and its influence on the simulation results is analyzed.

First Example

The example presented in Fig. 3.5 is investigated. Rod 1 with initial velocity v_{in} is hitting Rod 2 which is initially at rest. The impedance function of Rod 2 is constant ($\tilde{Z} = Z_0$). Rod 1 consists of two parts with impedance Z_0 and Z_1, respectively. The simulation is run with the parameters listed in Tab. 3.1. The corresponding results of the first

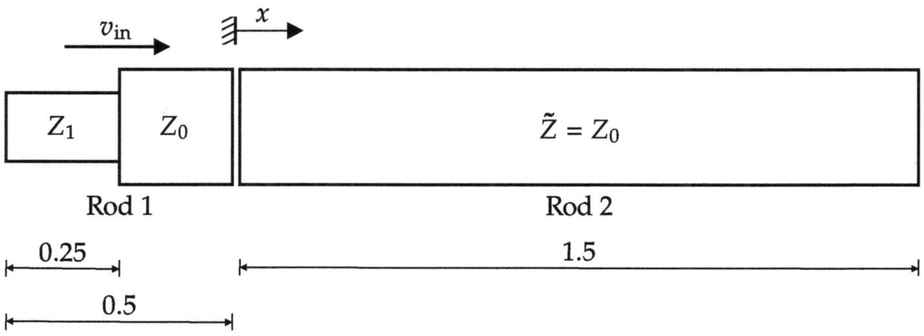

Figure 3.5: Setup of the first example. All dimensions in m.

Parameters		Number	Unit
Z_0	Impedance	$5.10 \cdot 10^4$	$\mathrm{N\,s\,m^{-1}}$
Z_1	Impedance	$2.87 \cdot 10^4$	$\mathrm{N\,s\,m^{-1}}$
v_{in}	Initial velocity of Rod 1	6.00	$\mathrm{m\,s^{-1}}$

Table 3.1: Parameters used for the first example.

example are presented in Fig. 3.6. The displacement (left) and the normal force (right) are depicted at three time instants. The transition between Rod 1 and Rod 2 is indicated by a vertical dashed line.

At the first time instant, compressive force waves are traveling in both rods. The part of the wavefront in Rod 1, which has been transmitted at the impedance jump of Rod 1, is traveling to the left. The reflected part is moving toward the transition between Rod 1 and Rod 2. The waves that have not been covered by the wavefronts yet are either traveling with initial velocity in Rod 1 or are at rest in Rod 2.

At the second time instant, the part of the incident wave in Rod 1, that has been reflected at the impedance jump, decreases the absolute value of the normal force in Rod 2 when it reaches the transition zone. Shortly after, when the incident wave in Rod 1 has traveled back and forth along Rod 1, the rods separate. Thus, both rods move independently which leads to discontinuous displacements of the rods at $x = 0$. After the separation, the normal forces periodically change in both rods.

At the third time instant, the incident waves in Rod 2 have been reflected at its free end. Therefore, the compressive force waves traveling to the right have been converted to tensile force waves traveling to the left. The tensile force waves are pulling the particles in positive $x-$direction. The displacements in Rod 2 are lowest in the region that has not been reached by the reflected tensile force waves yet. Before the compressive waves have been reflected at the free end, the displacements in this region have been largest (see second time instant). This procedure is permanently repeated and leads to continuously increasing displacements in Rod 2. Physically, the results are comprehensible as Rod 2 is moving in positive $x-$direction after the impact. The displacements and normal forces in Rod 1 may be interpreted analogously.

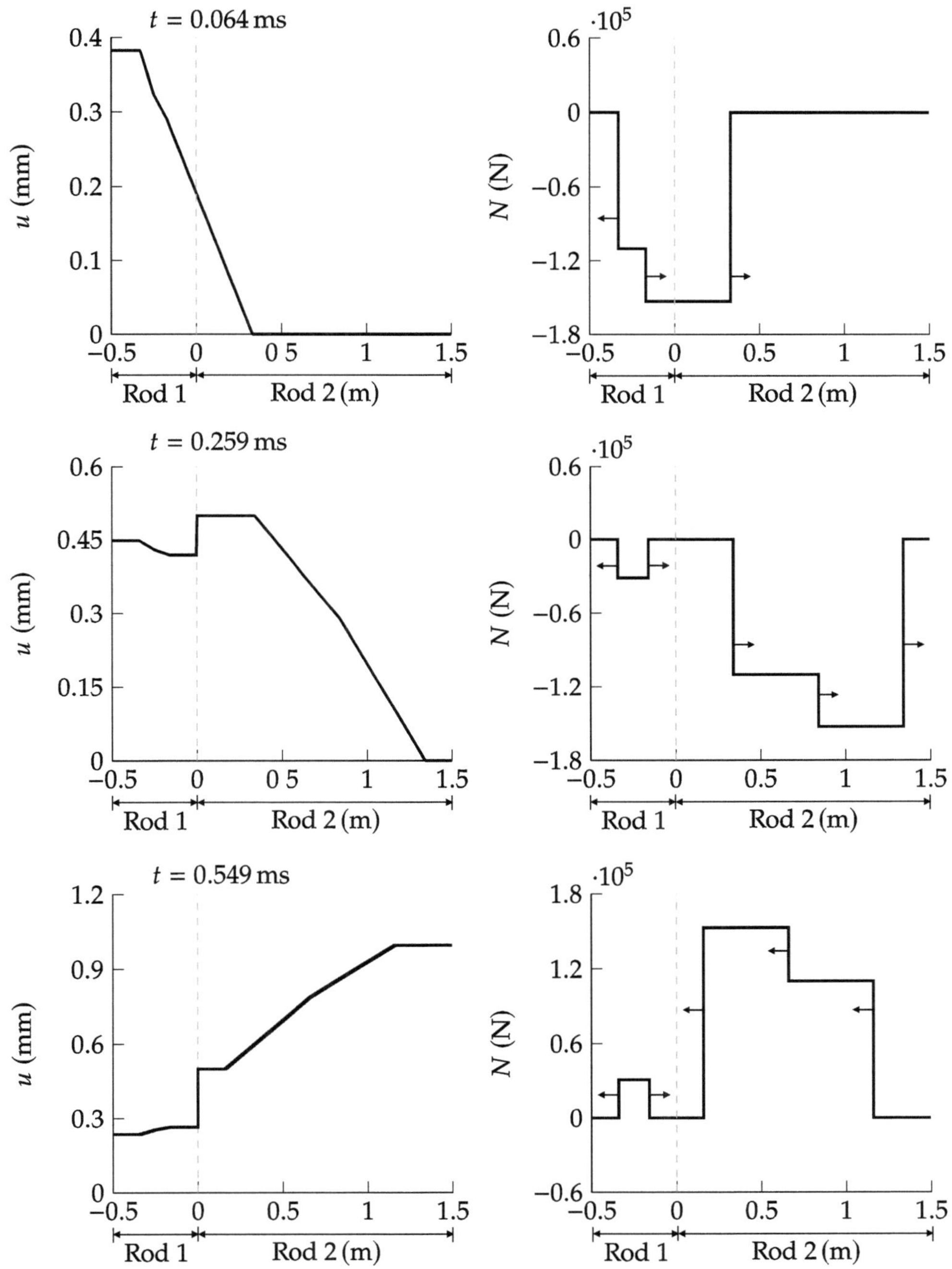

Figure 3.6: Results of the first example. The displacements (left) and the forces (right) are presented at $t = 0.064\,\text{ms}$ (first row), $t = 0.259\,\text{ms}$ (second row) and $t = 0.549\,\text{ms}$ (third row).

Complex Example

The setup depicted in Fig. 3.7 is investigated where Rod 1 with initial velocity v_{in} hits Rod 2 with initial velocity $\tilde{v}_{\text{in}}$. Both rods are supported by two plain bearings (see Fig. 3.7a). It is assumed that Rod 1 can move freely, whereas friction is considered at the plain bearings of Rod 2. The first plain bearing of Rod 2 is located between x_2 and x_3, the second is located between x_4 and x_5. With the mass

$$m = \sum_{j=1}^{\tilde{n}} \tilde{A}_j \tilde{\rho}_j \tilde{\ell}_j, \tag{3.21}$$

of Rod 2, the total friction force

$$F_{\text{R}} = \mu m \bar{g} \tag{3.22}$$

is calculated, where μ is the friction coefficient and $\bar{g}$ is the gravitational constant. If $\hat{n}$ is the total number of nodes k in contact with the plain bearings of Rod 2, then it is assumed that the total friction force is uniformly distributed

$$F_k^* = -\frac{F_{\text{R}}}{\hat{n}} \operatorname{sgn}\left(\frac{1}{2}(\tilde{v}_k + \tilde{v}_{k-1})\right). \tag{3.23}$$

The cross section area of Rod 1 is constant ($A = A_0$) as can be seen in Fig. 3.7a. The cross section area of Rod 2 linearly increases from A_0 to A_1 between x_1 and x_2. From x_2 to x_3, the cross section area is constant ($A = A_1$). Between x_3 and x_4, A is caluclated as a function of x

$$A(x) = A_1 - 0.3 A_0 \sin\left(\pi \frac{x - x_3}{x_4 - x_3}\right). \tag{3.24}$$

Subsequently, there is a constant section ($A = A_1$) between x_4 and x_5 followed by a linearly decreasing part from A_1 to A_0 between x_5 and x_6. In addition to the geometry variations, the material paramaters vary between x_3 and x_4 (see. Fig. 3.7b). This section may be interpreted as a rod part of different material ($E = 1.5\, E_0$, $\rho = 1.2\, \rho_0$) that is screwed together with the other rod parts of Rod 2. For example in rotary drilling, heavy weight drill rods are applied to reduce fatigue [16]. The geometry variations together with the material parameter variations lead to the impedance function depicted in Fig. 3.7c. Due to the changing material parameters in $x \in [x_3, x_4]$, the impedance function increases in this region which leads to jumps at $x = x_3$ and $x = x_4$.

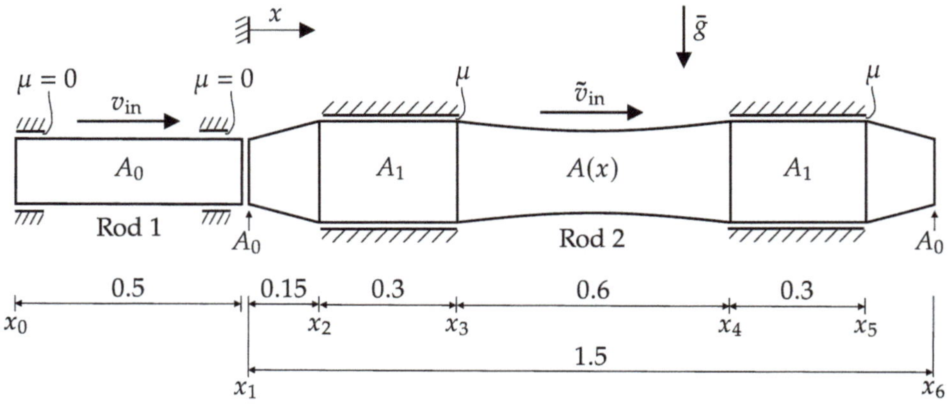

(a) Schematic sketch and dimensions (in m) of the investigated setup. Rod 1 with initial velocity $v_{in} > 0$ is hitting Rod 2 with initial velocity $\tilde{v}_{in} < 0$. Rod 1 can move freely ($\mu = 0$) in x–direcion, whereas a friction force ($\mu \neq 0$) is applied at the bearings of Rod 2.

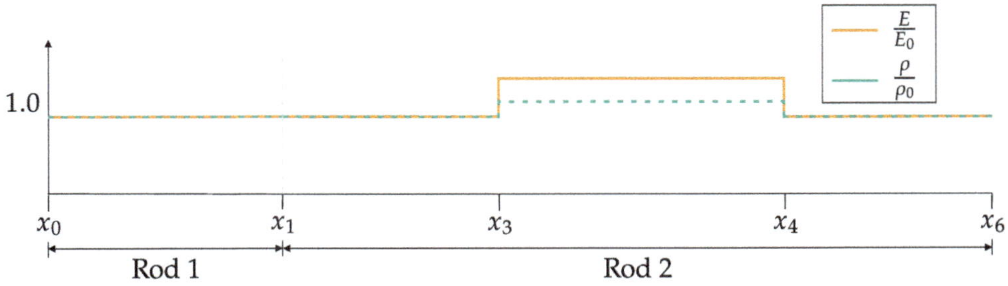

(b) Material parameters of the rods. Between x_3 and x_4, $E = 1.5\,E_0$ and $\rho = 1.2\,\rho_0$ applies. Rod 1 is homogeneous with $E = E_0$ and $\rho = \rho_0$.

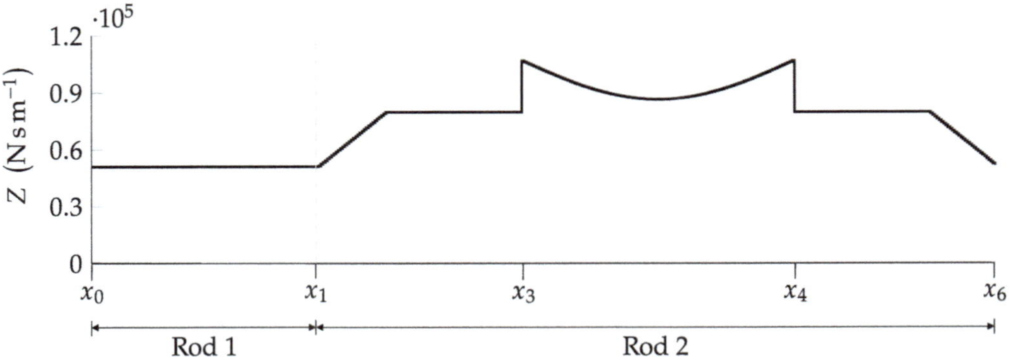

(c) Resulting impedance variations of the rods due to the geometry variations plotted in (a) and the material property variations described in (b).

Figure 3.7: Setup of the complex example.

Parameters		Number	Unit
A_0	Cross section area	$1.26 \cdot 10^{-3}$	m^2
A_1	Cross section area	$1.96 \cdot 10^{-3}$	m^2
E_0	Young's modulus	210	GPa
ρ_0	Density	$7.84 \cdot 10^3$	kg/m^3
v_{in}	Initial velocity of Rod 1	3.00	$m\,s^{-1}$
$\tilde{v}_{in}$	Initial velocity of Rod 2	-3.00	$m\,s^{-1}$
μ	Friction coefficient	0.3	–
$\bar{g}$	Gravitational constant	9.81	$N\,kg^{-1}$

Table 3.2: Parameters used for the complex example.

The simulation is run with the parameters listed in Tab. 3.2. In Fig. 3.8, the corresponding results for the displacement (left) and the normal force (right) are depicted at three time instants. The transition between Rod 1 and Rod 2 is indicated by a vertical dashed line. At the first time instant, the waves have propagated both into Rod 1 and Rod 2. The parts, that have not been covered by the wave front yet, travel with their initial velocities. At the second time instant, the wave front in Rod 1 is traveling toward the transition after it has been reflected at its free end. In Rod 2, two jumps in normal force may be observed. The first jump at $x \approx 0.14$ m results from the reflection of the incident wave front at the impedance jump at $x = x_3$. The transmitted part of the incident wave front is traveling at $x \approx 0.79$ m. Since the wave propagation speed is larger between x_3 and x_4 compared to the other parts, the transmitted part of the incident wave front has traveled further than the reflected part. At the last time instant, Rod 1 has separated from Rod 2. After the separation, free end boundary conditions are applied at the former hitting surface at $x = 0$. Therefore, the normal force at $x = 0$ is zero for both rods. Moreover, the rods can move independently in negative $x-$direction which explains the displacement jump at $x = 0$. The particles of Rod 2 at axial position $x > 1.38$ m have not been reached by the wave front which is why the absolute value of the displacement is maximum and the normal forces are close to zero in this region.

The normal forces are not exactly zero in this region, since the friction forces of the second bearing position have already influenced the normal forces. However, the influence of the friction forces on the results is very little. The reason is that the impact force is much larger compared to the friction forces. For the given example, the absolute value of the initial impact force is approximately 2250 times the absolute value of the total friction force. It can be concluded that if short travel times are considered, the friction forces are

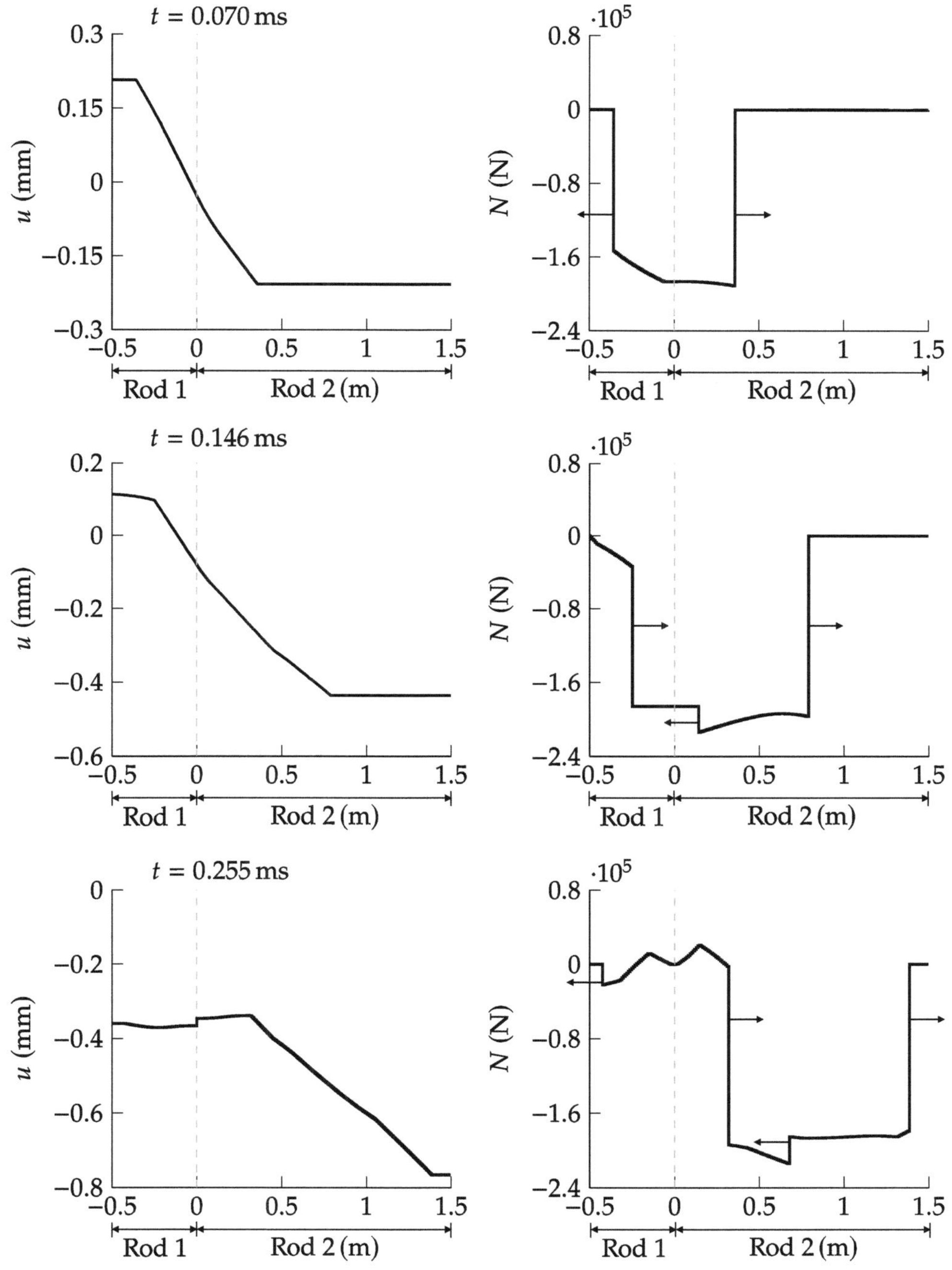

Figure 3.8: Results of the complex example. The displacements (left) and the forces (right) are presented at t = 0.070 ms (first row), t = 0.146 ms (second row) and t = 0.255 ms (third row).

negligible. In the long run, however, the friction forces significantly decrease the total energy E_{tot} of the system as can be seen in Fig. 3.9. The total energy

$$\begin{aligned} E_{tot} &= \frac{1}{2}\sum_{j=1}^{n} A_j \rho_j \ell_j v_j^2 + \frac{1}{2}\sum_{j=1}^{n} \frac{N_j^2}{E_j A_j}\ell_j + \frac{1}{2}\sum_{j=1}^{\tilde{n}} \tilde{A}_j \tilde{\rho}_j \tilde{\ell}_j \tilde{v}_j^2 + \frac{1}{2}\sum_{j=1}^{\tilde{n}} \frac{\tilde{N}_j^2}{\tilde{E}_j \tilde{A}_j}\tilde{\ell}_j \\ &= E_{kin,Rod\ 1} + E_{pot,Rod\ 1} + E_{kin,Rod\ 2} + E_{pot,Rod\ 2} \end{aligned} \tag{3.25}$$

consists of the kinetic and potential energy of both rods. For the evaluation, 100 iterations have been considered which means that the wave has traveled 100 times forth and back in Rod 2. Without friction, the system is conservative and the total energy remains constant. With friction, the total energy linearly decreases from its maximum value at the beginning to approximately 128 J at the end. Thus, in the long run, friction has to be considered.

The examples show that the presented wave propagation method is applicable to the longitudinal impact of rods with arbitrary material parameters and geometries. In addition, external forces that enable to model e.g. friction may be considered. The wave propagation method will be used as the basis for the recursive method, which is presented in the next chapter.

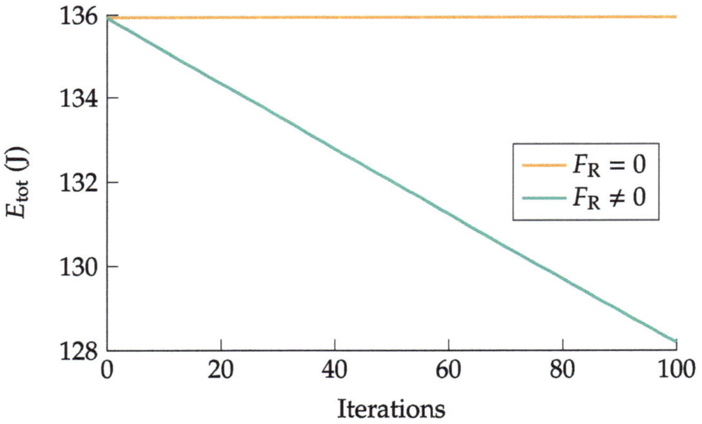

Figure 3.9: Comparison between the total energy E_{tot} with and without the consideration of friction in Rod 2. One iteration means that the wave has traveled back and forth in Rod 2.

3.1.4 Summary

In Fig. 3.10, the summary of the method is depicted. The chart is divided into three parts (application, preprocessing, simulation).

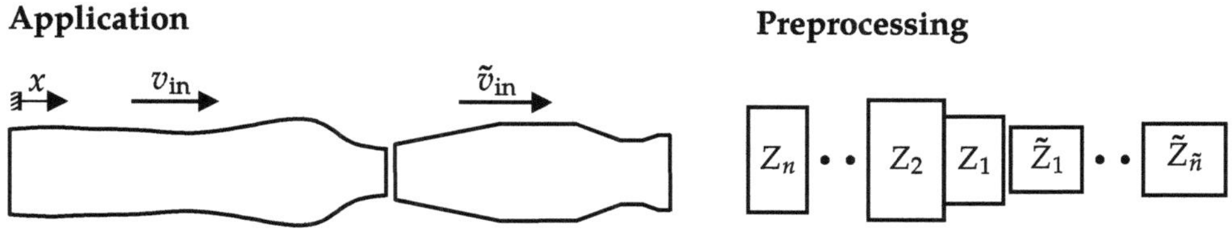

Simulation

$$t = 0 \quad : \quad N_j, v_j$$
$$F_j^*, N_j^+, N_j^-, v_j^+, v_j^- + \text{BC}$$
$$t = \Delta t \quad : \quad N_{j,\text{new}}, v_{j,\text{new}}$$

Figure 3.10: Summary of the wave propagation method.

A possible application of the presented method is the impact of non-uniform rods at initial velocities v_{in} and $\tilde{v}_{\text{in}}$, respectively. The rods are subdivided into piecewise constant impedances in the preprocessing step. Moreover, the element lengths are adjusted so that it takes one time step to travel through each element. Based on the initial displacements and velocities, the initial forces N_j and velocities v_j at $t = 0$ are determined. Afterward, the simulation starts with setting the current external forces F_j^*. Furthermore, the transition forces N_j^+, N_j^- and velocities v_j^+, v_j^- are determined and the current boundary conditions are set. The time step ends by calculating the new forces $N_{j,\text{new}}$ and velocities $v_{j,\text{new}}$ in equilibrium. This procedure is repeated until the desired simulation time is achieved.

The computational effort of the method is $O(n^*)$ with n^* being the total number of rod elements. This enables fast calculations using many elements.

3.2 Modal Expansion Method

As discussed in Sec. 2.2, analytical solutions exist for the longitudinal wave propagation in rods with varying cross section area. However, the possible cross sectional area variations are very limited. The aim of this section is to apply the modal expansion method to the problem given in Fig. 3.11. Thus, a reference solution is generated to

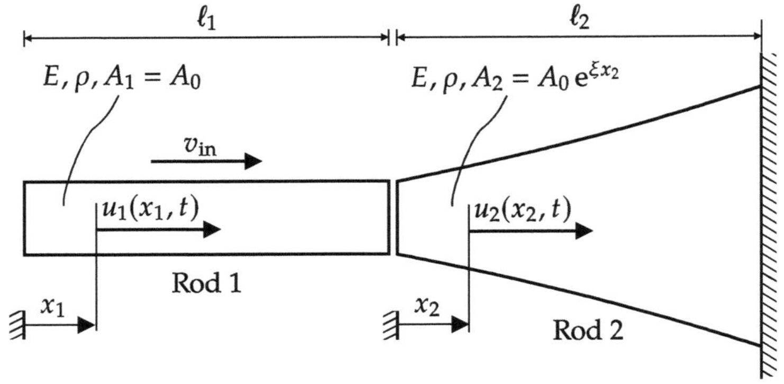

Figure 3.11: Investigated setup.

verify the wave propagation method. The setup consists of two homogeneous rods. Rod 1 is hitting Rod 2 with initial velocity v_{in}. The cross section area of Rod 2 increases exponentially and the displacements of the rods are denoted by $u_1(x_1, t)$ and $u_2(x_2, t)$, respectively.

Eigenfunctions

The eigenfunctions $U_n(x_n)$ of the approach

$$\begin{aligned} u_n(x_n, t) &= a_n(x)\, e^{i(k_n x_n - i\omega t)} \\ &= a_n(x)\, e^{ik_n x_n}\, e^{-i\omega t} = U_n(x_n)\, e^{-i\omega t}, \quad n = 1, 2 \end{aligned} \tag{3.26}$$

are indicated by substituting the related wave numbers and amplitudes of Tab. 2.1 into $U_n(x_n)$. This leads to the eigenfunctions

$$U_1(x_1) = C_1 \cos\left(\frac{\omega}{c_0} x_1\right) + C_2 \sin\left(\frac{\omega}{c_0} x_1\right), \tag{3.27}$$

$$U_2(x_2) = e^{\frac{-\xi x_2}{2}} \left[C_3 \cos\left(\sqrt{\left(\frac{\omega}{c_0}\right)^2 - \frac{\xi^2}{4}} x_2\right) + C_4 \sin\left(\sqrt{\left(\frac{\omega}{c_0}\right)^2 - \frac{\xi^2}{4}} x_2\right)\right] \tag{3.28}$$

of Rod 1 and Rod 2. The unknown coefficients $C_1 \ldots C_4$ are determined by evaluating the boundary and transition conditions. At the transition, displacement and force continuity are claimed. Moreover, the left end of Rod 1 is free whereas the right end of Rod 2 is fixed. The resulting equations

$$\left.\frac{\mathrm{d}U_1}{\mathrm{d}x_1}\right|_{x_1=0} = 0, \tag{3.29}$$

$$U_1(\ell_1) = U_2(0), \tag{3.30}$$

$$\left.\frac{\mathrm{d}U_1}{\mathrm{d}x_1}\right|_{x_1=\ell_1} = \left.\frac{\mathrm{d}U_2}{\mathrm{d}x_2}\right|_{x_2=0}, \tag{3.31}$$

$$U_2(\ell_2) = 0 \tag{3.32}$$

are written as a system of linear equations

$$\begin{pmatrix} 0 & \alpha & 0 & 0 \\ \cos(\alpha\ell_1) & \sin(\alpha\ell_1) & -1 & 0 \\ -\alpha\sin(\alpha\ell_1) & \alpha\cos(\alpha\ell_1) & \frac{\xi}{2} & -\beta \\ 0 & 0 & \cos(\beta\ell_2) & \sin(\beta\ell_2) \end{pmatrix} \begin{pmatrix} C_1 \\ C_2 \\ C_3 \\ C_4 \end{pmatrix} = \begin{pmatrix} 0 \\ 0 \\ 0 \\ 0 \end{pmatrix} \tag{3.33}$$

with

$$\alpha = \frac{\omega}{c_0}, \quad \beta = \sqrt{\alpha^2 - \left(\frac{\xi}{2}\right)^2}. \tag{3.34}$$

Evaluating the characteristic equation

$$\alpha\left(2\cos(\alpha\ell_1)\cos(\beta\ell_2)\beta + \cos(\alpha\ell_1)\sin(\beta\ell_2)\xi - 2\alpha\sin(\alpha\ell_1)\sin(\beta\ell_2)\right) = 0 \tag{3.35}$$

determines the infinite number of eigenfrequencies ω_k from which their corresponding coefficients

$$C_{1,k} = C_{3,k}\frac{1}{\cos\left(\frac{\omega_k}{c_0}\right)}, \tag{3.36}$$

$$C_{2,k} = 0, \tag{3.37}$$

$$C_{4,k} = -C_{3,k}\frac{1}{\tan\left(\left(\frac{\omega_k}{c_0}\right)^2 - \frac{1}{16}\right)} \tag{3.38}$$

are obtained. Thus, the eigenfunctions are determined and the approach is rewritten

$$\begin{aligned}\begin{Bmatrix} u_1(x_1,t) \\ u_2(x_2,t) \end{Bmatrix} &= \sum_{k=1}^{\infty} \begin{Bmatrix} U_{1,k}(x_1) \\ U_{2,k}(x_2) \end{Bmatrix} [C_{3,k}\tilde{C}_{5,k}\sin(\omega_k t) + C_{3,k}\tilde{C}_{6,k}\cos(\omega_k t)] \\ &= \sum_{k=1}^{\infty} \begin{Bmatrix} U_{1,k}(x_1) \\ U_{2,k}(x_2) \end{Bmatrix} [C_{5,k}\sin(\omega_k t) + C_{6,k}\cos(\omega_k t)].\end{aligned} \tag{3.39}$$

The remaining coefficients $C_{5,k}, C_{6,k}$ are obtained by evaluating the initial conditions.

Initial Conditions and Orthogonality Relation

For a PDE of the form

$$\vartheta(x)\frac{\partial^2 u(x,t)}{\partial t^2} + \mathcal{K}[u(x,t)] = 0, \quad x \in [0,\ell] \tag{3.40}$$

with linear differential operator $\mathcal{K}[\cdot]$, Hagedorn and DasGupta [36] have derived the orthogonality relation

$$\int_0^{\ell} \vartheta(x)U_j(x)U_k(x)\,\mathrm{d}x = B_k\delta_{jk}, \tag{3.41}$$

where B_k is a constant. Applying this to the given problem of two impacting rods with varying cross section area and constant material parameters leads to

$$\vartheta_i(x_i) = \rho A_i(x_i), \quad \mathcal{K}_i[\cdot] = -\frac{\partial}{\partial x_i}\left(EA_i(x_i)\frac{\partial}{\partial x_i}\right), \quad i = 1,2 \tag{3.42}$$

and the orthogonality relation is

$$\begin{aligned}&\int_0^{\ell_1} A_1(x_1)U_{1,j}(x_1)U_{1,k}(x_1)\,\mathrm{d}x_1 \\ &+ \int_0^{\ell_2} A_2(x_2)U_{2,j}(x_2)U_{2,k}(x_2)\,\mathrm{d}x_2 = \begin{cases} B_k, & j = k \\ 0, & \text{else.}\end{cases}\end{aligned} \tag{3.43}$$

Finally, the initial conditions for the displacements

$$\begin{Bmatrix} u_1(x_1,0) \\ u_2(x_2,0) \end{Bmatrix} = \sum_{k=1}^{\infty} \begin{Bmatrix} U_{1,k}(x_1) \\ U_{2,k}(x_2) \end{Bmatrix} C_{6,k} \tag{3.44}$$

and for the velocities

$$\begin{Bmatrix} v_1(x_1,0) \\ v_2(x_2,0) \end{Bmatrix} = \sum_{k=1}^{\infty} \begin{Bmatrix} U_{1,k}(x_1) \\ U_{2,k}(x_2) \end{Bmatrix} C_{5,k}\omega_k \tag{3.45}$$

are evaluated and the orthogonality relation is applied. Thus, the remaining coefficients

$$C_{5,k} = \frac{1}{B_k\omega_k}\left[\int_0^{\ell_1} v_1(x_1,0)A_1(x_1)U_{1,k}(x_1)\,\mathrm{d}x_1 + \int_0^{\ell_2} v_2(x_2,0)A_2(x_2)U_{2,k}(x_2)\,\mathrm{d}x_2\right] \tag{3.46}$$

and

$$C_{6,k} = \frac{1}{B_k}\left[\int_0^{\ell_1} u_1(x_1,0)A_1(x_1)U_{1,k}(x_1)\,\mathrm{d}x_1 + \int_0^{\ell_2} u_2(x_2,0)A_2(x_2)U_{2,k}(x_2)\,\mathrm{d}x_2\right] \tag{3.47}$$

are obtained.

3.3 Comparison between Wave Propagation and Modal Expansion Method

For piecewise constant impedance functions of the rods, the wave propagation method yields exact solutions up to a certain time as a sum of a finite number of terms (closed-form). As compared to this, the modal expansion method leads to approximate solutions for all time if a finite number of terms is used. Both methods are exact for all time if one makes use of an infinite number of terms. Therefore, if it is only of interest to determine the solution up to a certain time, the wave propagation method is more accurate. These statements are demonstrated by applying both methods to the example depicted in Fig. 3.11 and comparing their results. Rod 1 is hitting Rod 2 at the initial velocity v_{in}. Moreover, the initial displacements $u_1(x_1,0)$ and $u_2(x_2,0)$ are zero. The other parameter values are listed in Tab. 3.3.

Parameters		Number	Unit
E	Young's modulus	210	GPa
ρ	Density	$7.84 \cdot 10^3$	kg/m^3
v_{in}	Initial velocity of Rod 1	6.00	m s^{-1}
ℓ_1	Length of Rod 1	1.00	m
ℓ_2	Length of Rod 2	1.00	m
A_0	Cross section area at $x = 0$	$1.96 \cdot 10^{-3}$	m^2
ξ	Exponent coefficient of Rod 2	$\frac{1}{2}$	m^{-1}

Table 3.3: Parameters of the example presented in Fig. 3.11.

Figure 3.12 shows the corresponding results of the displacements and forces at three different times. The hit takes place at $x = 0$ which is highlighted by a vertical dashed line. To the left of the dashed line, the results of Rod 1 are plotted whereas the results of Rod 2 are depicted on the right side. The rods stick together as long as the force at the transition is compressional. This condition always applies to the presented time steps.

At the first time, the incident waves have propagated 0.75 m in both rods. The part of the rods that is further away than 0.75 m from the hitting surface is not influenced by the impact yet. Thus, these elements are either at rest (Rod 2) or traveling at initial velocity (Rod 1) as can be seen on the left graph of the first row in Fig. 3.12. Furthermore, there is no significant deviation between the displacement results. These curves match significantly better than the force curves. The reason is that the force curves are not continuous. In the zoom box, the overshooting of the modal expansion method result at the jump is highlighted. This behavior is characteristic for eigenfunction series and is well known in literature as the Gibbs phenomenon. The forces in the parts of the rods, that are not covered by their wavefronts yet, are still zero. However, the oscillations of the modal expansion method results suggest that the wave already traveled further than it actually did. Apart from the region close to the discontinuities, the force curves are in very good accordance.

The continuous increase in cross section area of Rod 2 leads to continuously increasing forces according to amount in Rod 2 and at the transition. Since the cross section area of Rod 1 is constant, the wave shape does not change while propagating along Rod 1. Thus, the impact force at $t = 0$ conforms with the wavefront of Rod 1.

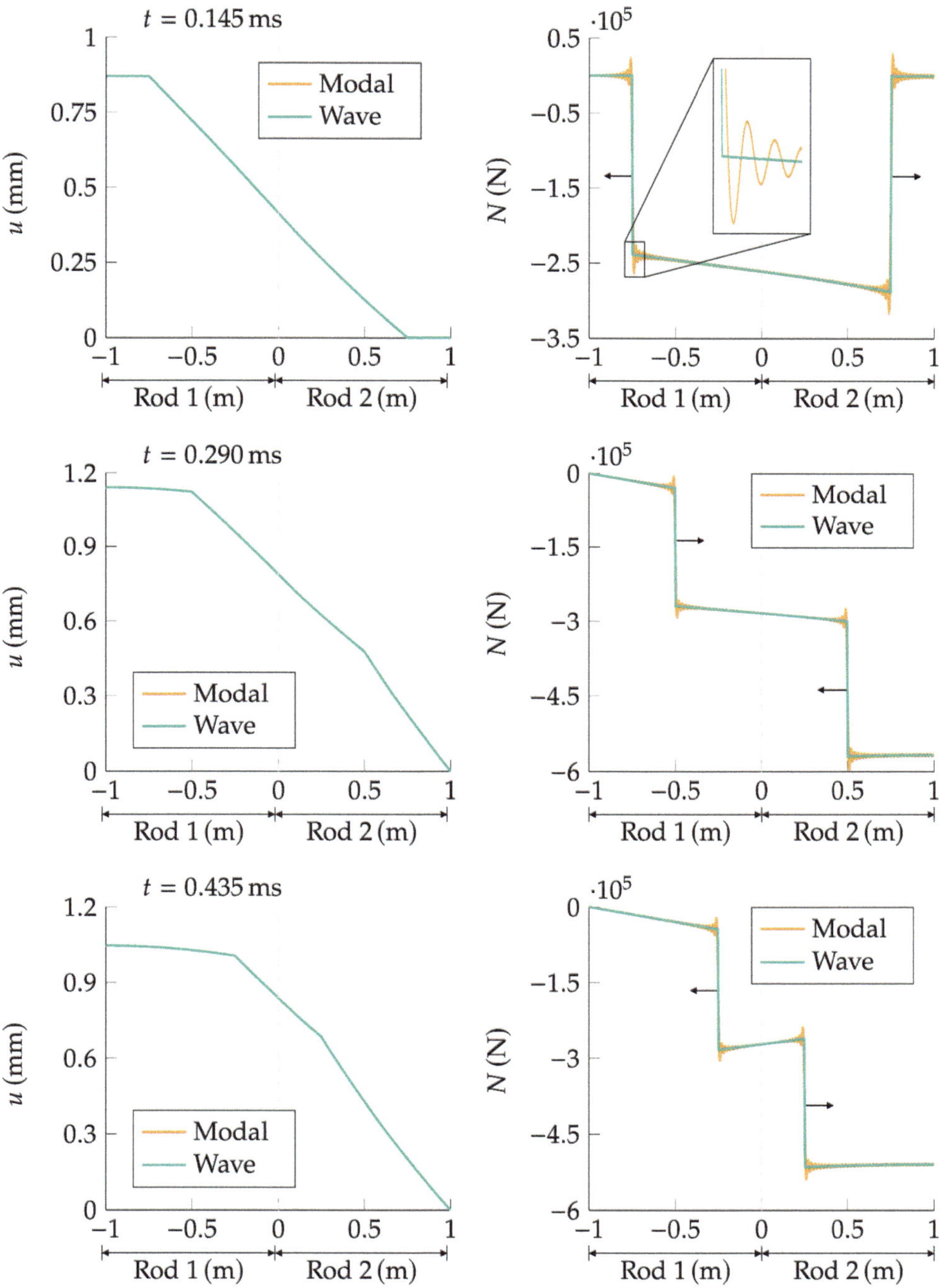

Figure 3.12: Comparison of results obtained with the modal expansion method versus results obtained with the wave propagation method. The displacements (left) and the forces (right) are presented at $t = 0.145$ ms (first row), $t = 0.290$ ms (second row) and $t = 0.435$ ms (third row).

At the second time, the waves have been reflected at their respective boundaries and travel back toward the transition. The free end on the left side of Rod 1 leads to zero force at $x = -1$, whereas the fixed end of Rod 2 results in zero displacement at $x = 1$. When the wavefront of Rod 1 reaches its left end, the maximum displacement is achieved. Afterward, the reflected tensile force wave reduces the maximum deflection.

At the third time, the wavefronts have met at the transition and are traveling back to their ends. This time, the reflected waves yield lower force magnitudes in Rod 2 and greater magnitudes in Rod 1. The force jumps always occur at the curve bends of the displacements.

In summary, the comparison of the results obtained with the two methods shows very good accordance. The deviations at the region close to the force jumps are explainable with the Gibbs phenomenon. It is disadvantageous for the modal expansion method that a large number of eigenfunctions (2500 for the presented example) have to be considered to plot the jumps. Moreover, the modal expansion method will always yield approximate solutions as an infinite number of eigenfunctions can not be considered. It can be concluded that for impact problems which include force jumps, the wave propagation method is preferable. However, most importantly, the wave propagation method is verified by the modal expansion method.

4 Impacting Rods: Inverse Problem

Inverse problems occur in many practical applications. A well-known example is the computer tomography where by detecting an attenuated X-Ray, the spatial X-Ray absorption is concluded. This enables to get non-destructive information about the scanned structure. The purpose of the presented method is to determine the impedance function of Rod 1 which leads to the desired force over time at an arbitrary position of Rod 2. To solve the inverse problem, the impedance function of Rod 2 and the initial and boundary conditions have to be known (see Fig. 4.1). For the solution of the direct problem, only the impedance functions of the rods and the initial and boundary conditions have to be given to calculate the force and displacement distribution in both rods.

	Given	Calculated
Direct	Impedance functions of the rods + Initial and boundary conditions	E.g. force/displacement distribution in both rods
Inverse	Impedance function of Rod 2 + Initial and boundary conditions + Force over time at one position of Rod 2	Impedance function of Rod 1

Figure 4.1: Overview of given and calculated values of the direct and inverse problem.

The following chapter is structured as follows: Initially, the recursive method (RM) is presented. The term reursive refers to the way how the inverse problem is solved. The implementation of the main idea is presented and a condition of existence is derived. It is shown how the impedance changes correlate with the probability that a solution of the inverse problem exists. Moreover, it is shown that a prescribed force at an arbitrary position of the second rod may be transformed to the corresponding impact force. At the end of the chapter, several examples are presented, including the application in percussion drilling.

4.1 Recursive Method

The RM is based on the wave propagation method presented in Sec. 3.1. Without loss of generality, the hit takes place at $t = 0$ and Rod 1 is the rod with unknown impedance. In addition to the assumptions that are valid for the wave propagation method, some prerequisites have to be met to apply the method.

These **prerequisites** are:

- Rod 1 is stress free just before the impact.
- Before the impact, all particles of Rod 1 are moving with the same velocity.
- No external forces are acting on the nodes of Rod 1.

For Rod 2, there are no prerequisites.

In Fig. 4.2a, a schematic sketch of the elements close to the impact zone is depicted. The rods hit each other at initial velocities v_{in} of Rod 1 and $\tilde{v}_{\text{in}}$ of Rod 2. Moreover, the impedances of Rod 2 are given. The impedances of Rod 1 are determined according to the predefined impact force (target force) which is illustrated in Fig. 4.2b. The main idea of the method is to utilize that each element of Rod 1 influences the impact force N_{I} at different moments for the first time. Immediately after the impact, the influence of the first element of Rod 1 can be observed in Fig. 4.2b. Two time steps later, the influence of the second element becomes obvious. If the first $q-1$ elements in Fig. 4.2 are set to perfectly match the target values, the question remains how the qth element has to be set to continuously match the target force. Initially, the simulation is carried out with the same impedance as the preceding element ($Z_q = Z_{q-1}$). The difference between the target force N_q (dashed line) and the force $\bar{N}_q$ obtained with equal impedances is characterized by ΔN_q. The occurring force difference is balanced by an adjustment of the impedance Z_q. The only possibility that the impedance Z_q influences the impact force N_{I} at the considered time is if the initial impact force N_1 is directly transmitted to the transition between element $q-1$ and q where it is reflected and directly transmitted back until it reaches the hitting surface. The mathematical implementation of the idea is presented in the following section.

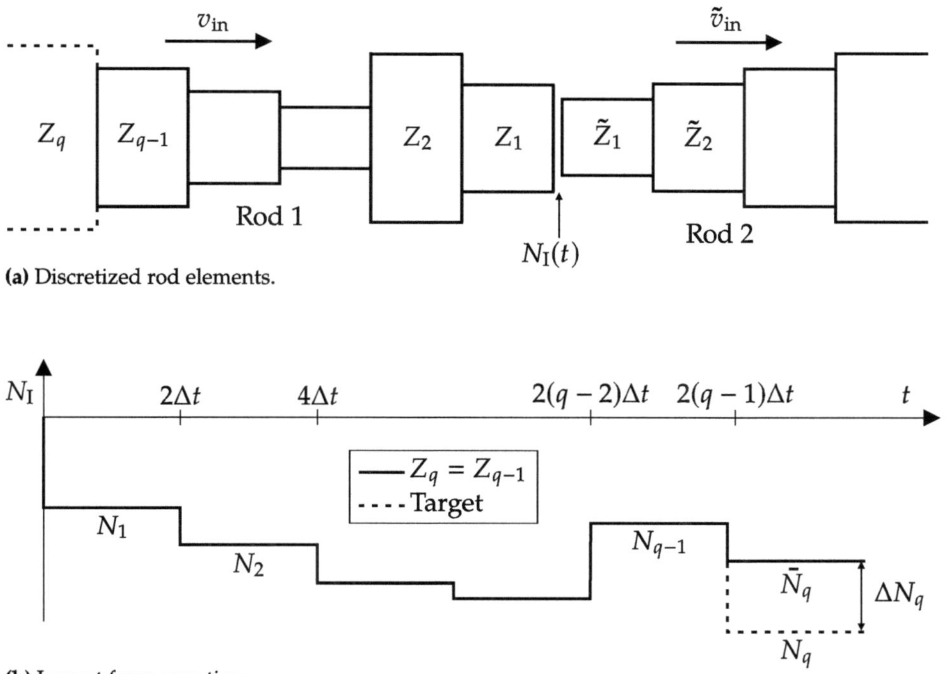

(a) Discretized rod elements.

(b) Impact force over time.

Figure 4.2: Idea of the recursive method.

As the waves have to travel back and forth at each element before they reach the hitting surface, it is only possible to change N_I every second time step by an impedance change of Rod 1. Therefore, the target force, the boundary force and the initial conditions are assumed to change only every second time step. However, this is not a serious limitation as by doubling the number of elements, the same discretization accuracy is obtained.

4.1.1 Prescribed Impact Force

Initially, the formula to calculate the impedance of the first element is derived. Subsequently, the formula to recursively determine all the other elements of Rod 1 is presented. There is a different treatment of the first and all the other elements of Rod 1. The impedance of the first element is set to determine the appropriate initial impact force whereas for all the other elements, the initial impact force is given.

First element

The initial impact force

$$N_1 = \frac{2Z_1}{Z_1 + \tilde{Z}_1} \frac{(\tilde{v}_{\text{in}} - v_{\text{in}})\tilde{Z}_1}{2} = T_{\tilde{1},1} N_0 \tag{4.1}$$

is obtained by inserting the initial conditions for the first element of Rod 1 and Rod 2 in Eq. (3.6). It is composed of the transmission factor

$$T_{\tilde{1},1} = \frac{2Z_1}{Z_1 + \tilde{Z}_1} \tag{4.2}$$

from the first element of Rod 2 to the first element of Rod 1 (see Eq. (2.39)) times the impact force

$$N_0 = -\frac{(v_{\text{in}} - \tilde{v}_{\text{in}})\tilde{Z}_1}{2}. \tag{4.3}$$

Rearranging Eq. (4.1) and solving for the unknown impedance Z_1 yields

$$Z_1 = -\frac{N_1 \tilde{Z}_1}{N_1 + (v_{\text{in}} - \tilde{v}_{\text{in}})\tilde{Z}_1}. \tag{4.4}$$

Other elements

The remaining elements (second until nth element) are determined by implementing the idea that every element influences the impact force at different moments for the first time. In Fig. 4.2, the first $q-1$ impedances are set to perfectly match the target force. The impedance of the next element q is determined by first setting $Z_q = Z_{q-1}$ and determining the impact force $\bar{N}_q$. Subsequently, the force difference ΔN_q between target force N_q and $\bar{N}_q$ is balanced by

$$\begin{aligned} \Delta N_q = N_q - \bar{N}_q &= N_1\, T_{1,2}\, T_{2,3} \dots T_{q-2,q-1}\, R_{q-1,q}\, T_{q-1,q-2} \dots T_{3,2}\, T_{2,1}\, T_{1,\tilde{1}} \\ &= N_0\, T_{\tilde{1},1}\, T_{1,2}\, T_{2,3} \dots T_{q-2,q-1}\, T_{q-1,q-2} \dots T_{3,2}\, T_{2,1}\, T_{1,\tilde{1}}\, R_{q-1,q} \\ &= \left(T_{\tilde{1},1}\, T_{1,\tilde{1}}\right) (T_{1,2}\, T_{2,1}) \dots \left(T_{q-2,q-1}\, T_{q-1,q-2}\right) R_{q-1,q} N_0, \end{aligned} \tag{4.5}$$

where $T_{i,j}, T_{j,i}$ are the transmission factors from element i to element j respectively from element j to element i and $R_{q-1,q}$ is the reflection factor from element $q-1$ to element q. There is only one wave train that influences the impact force due to the impedance change of Z_q: The initial impact force N_1 which is directly transmitted to the $(q-1)$th element ($T_{1,2} \dots T_{q-2,q-1}$) where it is reflected at the transition to the qth element ($R_{q-1,q}$) and transmitted back to the impact surface ($T_{q-1,q-2} \dots T_{1,\tilde{1}}$). Since

$$T_{i,j}\, T_{j,i} = \frac{4Z_i Z_j}{(Z_i + Z_j)^2} = K_{i,j} = K_{j,i} \tag{4.6}$$

holds, the force difference can be written in a compact form

$$\begin{aligned} \Delta N_q &= K_{\tilde{1},1}\, K_{1,2} \dots K_{q-2,q-1}\, R_{q-1,q} N_0 \\ &= K_q R_{q-1,q} N_0 \\ &= K_q \frac{Z_q - Z_{q-1}}{Z_q + Z_{q-1}} N_0, \end{aligned} \tag{4.7}$$

where

$$K_q = K_{\tilde{1},1} \prod_{r=1}^{q-2} K_{r,r+1} = K_{\tilde{1},1} \prod_{r=1}^{q-2} \frac{4Z_r Z_{r+1}}{(Z_r + Z_{r+1})^2}. \tag{4.8}$$

Rearranging Eq. (4.7) and solving for Z_q leads to the final equation for the unknown impedance

$$Z_q = Z_{q-1} \frac{K_q\, N_0 + \Delta N_q}{K_q\, N_0 - \Delta N_q}. \tag{4.9}$$

After determining Z_q, the procedure repeats for the next element until all elements of Rod 1 are calculated recursively.

4.1.2 Transformation of the Prescribed Force

If the force N_p is prescribed at an arbitrary position p of Rod 2, it is possible to transform N_p to the corresponding impact force N_I (see Fig. 4.3a). The presented procedure makes use of the linearity of the problem. Thus, the impact force is reconstructed by a sum of weighted time-shifted responses to a rectangular excitation. The prerequisites of the RM and the piecewise constant modeling of the rod's impedances lead to forces that only vary every second time step. The prescribed force $N_p\,(t)$ at position p is defined by its factors Y_j

$$N_p\,(t) = Y_j,\;\; t_\mathrm{dt} + 2(j-1)\Delta t \leq t < t_\mathrm{dt} + 2j\Delta t,\;\; j = 1 \dots n \tag{4.10}$$

as can be seen in Fig. 4.3b.

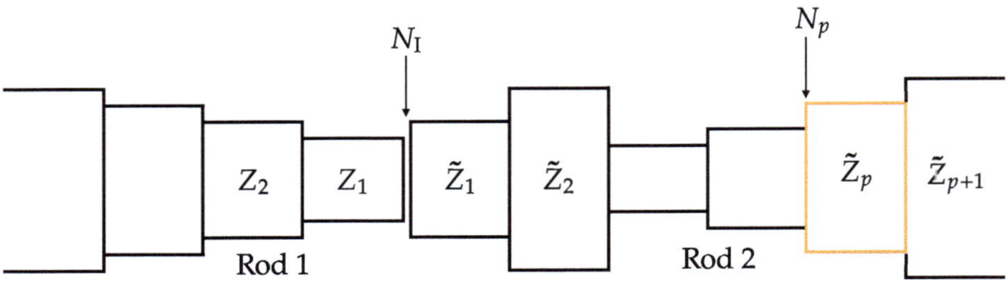

(a) Overview of the problem.

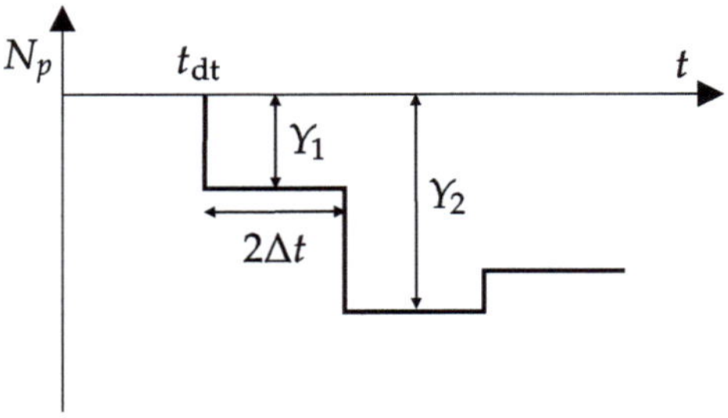

(b) Discretized force N_p.

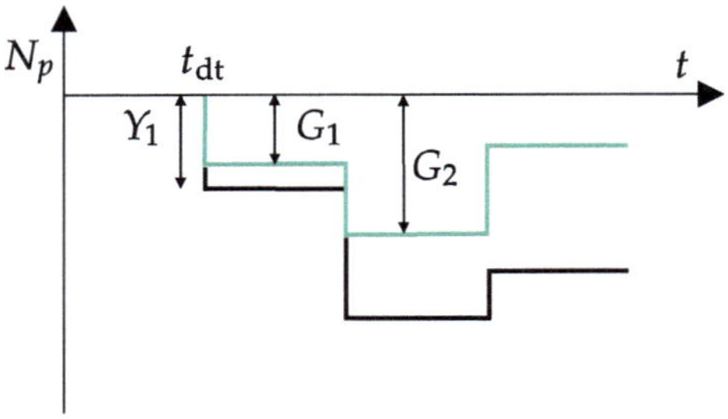

(c) Rectangular response (green).

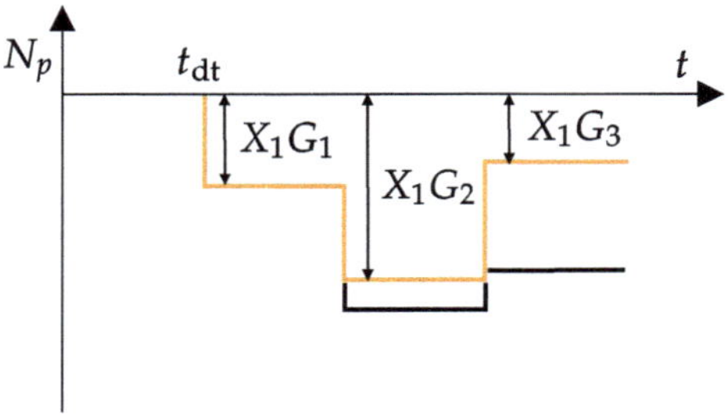

(d) Weighted rectangular response (orange). X_1 is set to perfectly match the first step.

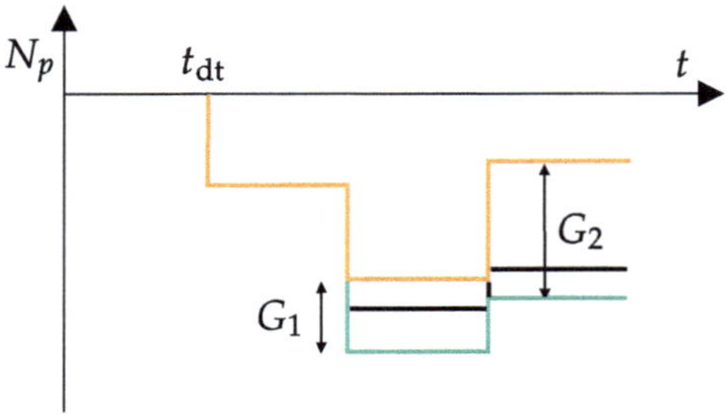

(e) Weighted rectangular response (orange) plus the time shifted rectangular response (green).

Figure 4.3: Recursive determination of the factors X_j of the impact force N_I.

The aim is to determine the factors X_j

$$N_\mathrm{I}(t) = X_j, \;\; 2(j-1)\Delta t \le t < 2j\Delta t, \;\; j = 1\dots n \tag{4.11}$$

of the unknown force N_I which generate N_p. To this end, a rectangular function

$$N_\mathrm{I}(t) = \mathcal{H}(t) - \mathcal{H}(t - 2\Delta t), \tag{4.12}$$

with $\mathcal{H}(\cdot)$ being the Heaviside function, is applied. Its response at position p

$$G(t) = G_j, \ \ t_{\mathrm{dt}} + 2(j-1)\Delta t \leq t < t_{\mathrm{dt}} + 2j\Delta t, \ \ j = 1 \ldots n \tag{4.13}$$

is called rectangular response. In Fig. 4.3c, the rectangular response together with the desired force is depicted. The rectangular response is weighted so that it matches the first step of N_p (see Fig. 4.3d). Therefore, the first factor is determined as

$$X_1 = \frac{Y_1}{G_1}. \tag{4.14}$$

All the other time-shifted rectangular responses arrive later and are not able to influence the first step. In addition to the first rectangular function, a second rectangular function

$$N_{\mathrm{I}}(t) = \mathcal{H}(t - 2\Delta t) - \mathcal{H}(t - 4\Delta t), \tag{4.15}$$

which is time-shifted by $2\Delta t$, is applied (see Fig. 4.3e). The second rectangular response is weighted to match the second step. This enables the calculation of the second factor

$$X_2 = \frac{1}{G_1}\left(Y_2 - X_1 G_2\right). \tag{4.16}$$

The procedure is repeated until all factors X_j are determined recursively. In general, the formula for the determination of the jth factor is

$$X_j = \frac{1}{G_1}\left(Y_j - \sum_{k=1}^{j-1} X_k G_{j+1-k}\right). \tag{4.17}$$

4.1.3 Condition of Existence

It is not always possible to generate a solution of the inverse problem for any given force over time. Physical solutions only exist if the obtained impedances in Eq. (4.9) are positive, i. e.

$$Z_q > 0. \tag{4.18}$$

Therefore, the following condition

$$\frac{K_q N_0 + \Delta N_q}{K_q N_0 - \Delta N_q} > 0 \tag{4.19}$$

holds which is equivalent to

$$\left|\Delta N_q\right| < \left|K_q\, N_0\right|. \tag{4.20}$$

The force N_0 is problem specific but independent of the determined element. By rewriting the factor K_q as

$$K_q = K_{\tilde{1},1} \prod_{r=1}^{q-2} K_{r,r+1} = K_{\tilde{1},1} \prod_{r=1}^{q-2} \left[1 - \left(\frac{Z_r - Z_{r+1}}{Z_r + Z_{r+1}}\right)^2\right] \leq 1, \tag{4.21}$$

it becomes obvious that the absolute value of the force reserve $\left|K_q\, N_0\right|$ decreases if the consecutive impedances change. This means that the larger changes in impedance are, the lower are the force reserves. Therefore, lower force differences $\left|\Delta N_q\right|$ may be compensated as condition (4.20) has to hold. Physically, the factor K_q may be interpreted as the fraction of transmitted energy through the impactor. The more impedance steps are passed and the higher the jumps are, the more internal reflections occur whose energy cannot be used to compensate the force differences.

In Fig. 4.4, the factors $K_{r,r+1}$ and K_n are analyzed for a constant change in impedance, that is $Z_{r+1} = mZ_r$. The factor

$$K_n = \prod_{r=1}^{n} \left[1 - \left(\frac{Z_r - Z_{r+1}}{Z_r + Z_{r+1}}\right)^2\right] = \left[1 - \left(\frac{1-m}{1+m}\right)^2\right]^n \tag{4.22}$$

describes the fraction of total transmitted energy for n transitions and

$$K_{r,r+1} = 1 - \left(\frac{1-m}{1+m}\right)^2 \tag{4.23}$$

describes the fraction of total transmitted energy at one transition. As the fraction of total transmitted energy at one transition is independent of whether the wave is first transmitted from the element with lower impedance to the element with larger impedance or reverse, $K_{r,r+1}(m) = K_{r,r+1}\left(\frac{1}{m}\right)$ applies. Therefore, only values $m \geq 1$ are plotted. For constant impedances ($m = 1$), all the energy is transmitted (see Fig. 4.4a). The larger the changes in impedance are, the lower is the fraction of transmitted energy. In Fig. 4.4b, it is shown that large jumps in impedance lead to disproportionally large decreases in K_n. For example, one jump in impedance with $m = 2$ leads to larger

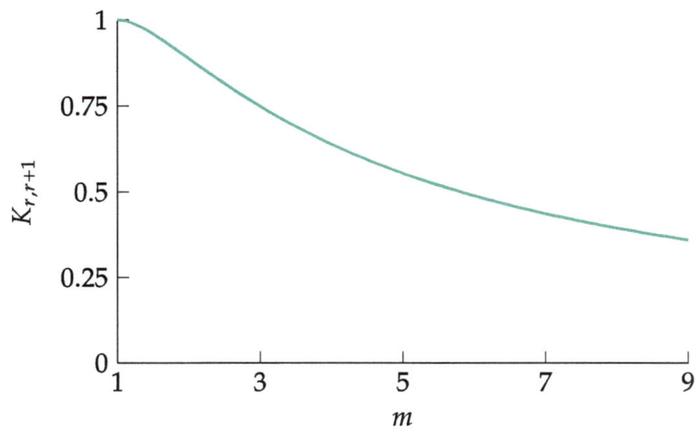

(a) $K_{r,r+1}$ for $Z_{r+1} = mZ_r$.

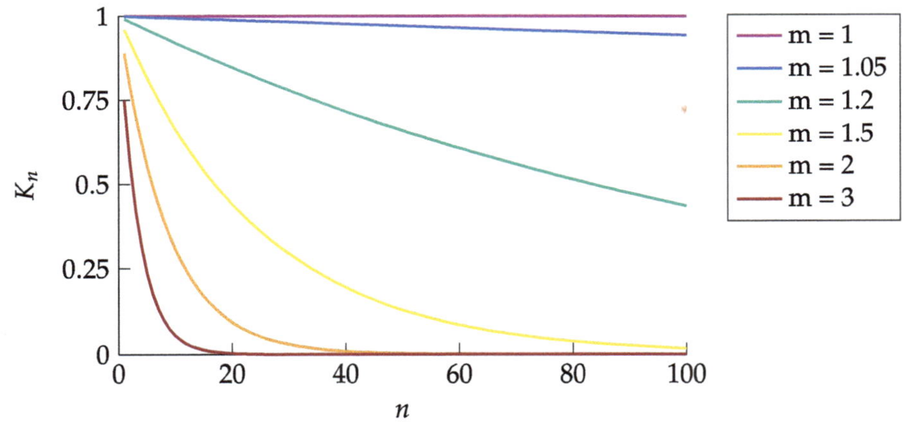

(b) K_n for $Z_{r+1} = mZ_r$.

Figure 4.4: Analysis of the continuously decreasing force reserve.

decreases in K_n than 100 jumps with $m = 1.05$ although the initial impedance value only doubles in the first case and approximately 132-folds in the second case.

4.1.4 Summary

The presented method applies to many inverse impact problems. The position, where the force over time is given, can be chosen arbitrarily if the prerequisites are met. It is only possible to prescribe the force over time for a time span which is twice the travel time of the waves through Rod 1. Afterward, there are no elements left to shape the waves. By applying the RM, the two main questions concerning inverse problems are answered: Firstly, there is a solution of the inverse problem (**existence**) if the condition in Eq. (4.20) holds for every element of Rod 1. Moreover, in case the condition is violated, it is possible to assess the maximum length of Rod 1 for which a solution of the inverse problem still exists. Secondly, the solution is unique (**uniqueness**) in terms of impedance. Another important aspect that is frequently discussed in inverse problems is the stability of the results with respect to disturbed measurements. However, as the prescribed force in the presented impacting rod problem is desired and not measured, there are no deviations which could lead to instabilities.

The calculation time of the algorithm (see Tab. 4.1) is $O(n^{*2})$ with n^* being the total number of rod elements. On a standard laptop, the solution is obtained within 1 s if less than 400 elements are chosen.

Algorithm: Recursive Method

Calculate Z_1.
FOR q from 2 to n
    Run simulation with $Z_q = Z_{q-1}$.
    Determine ΔN_q and K_q.
    IF $|\Delta N_q| < |K_q\, N_0|$
        Calculate Z_q.
    ELSE
        Solution of the inverse problem does not exist.
    END
END

Table 4.1: Algorithm of the RM.

4.2 Examples

The working principle of the recursive method is illustrated by three examples. The setups for the first two academic examples only differ in the prescribed impact force. For an exponentially decreasing impact force, a solution of the inverse problem exists. Moreover, for this particular setup, an exact reference solution exists which is compared with the corresponding results obtained by applying the RM. With the second academic example, it is shown that it is only possible to calculate a solution up to a certain time if a linear increasing impact force is generated. The last example is devoted to the practical application of percussion drilling. Thus, the impedance of the impacting piston is determined to generate a desired incident force in the rock.

4.2.1 Academic Examples

The setup illustrated in Fig. 4.5 is considered where a semi-infinite rod (Rod 2) with constant impedance $\tilde{Z}$ is given. Rod 2 is at rest at the beginning ($\tilde{v}_{\mathrm{in}} = 0$). Moreover, the velocity v_{in} of Rod 1 is known. The aim is to determine the unknown impedance function $Z(s)$ of Rod 1 which generates a prescribed impact force $N_{\mathrm{I}}(t)$. The parameters

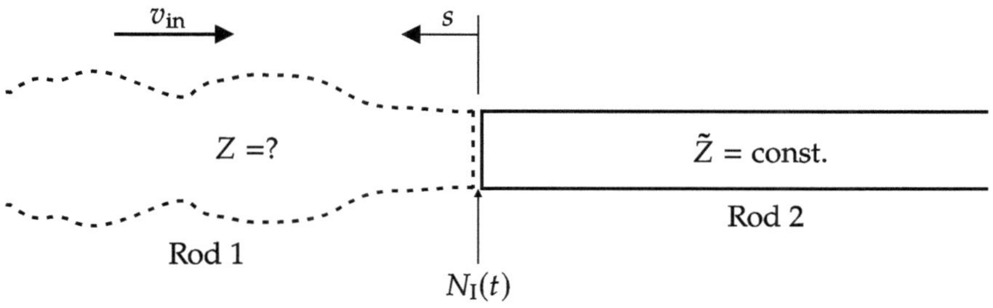

Figure 4.5: Setup of the academic examples. The aim is to determine the impedance Z which generates the impact force $N_{\mathrm{I}}(t)$ for given velocity v_{in} and given impedance $\tilde{Z}$.

listed in Tab. 4.2 are chosen for both examples.

Parameters		Number	Unit
$\tilde{Z}$	Impedance of Rod 2	79.7	$\mathrm{kN\,s\,m^{-1}}$
$\hat{N}$	Impact force at $t = 0$	239	kN
t_0	Time constant	0.383	ms
Δt	Time step	1.93	µs
n	Number of elements of Rod 1	100	—
v_{in}	Velocity of Rod 1	6.00	$\mathrm{m\,s^{-1}}$

Table 4.2: Parameters for the academic examples.

Example 1: Exponentially Decreasing Impact Force

Initially, an exponentially decreasing impact force

$$N_{\mathrm{I}}(t) = -\hat{N}\,\mathrm{e}^{\left(-\frac{t}{t_0}\right)} \tag{4.24}$$

with arbitrary time constant t_0 is given. LUNDBERG and LESSER [47] have introduced the traveling time of the wavefront starting from the impact surface traveling to position s with

$$\theta(s) = \int_0^s \frac{1}{c(s')}\,\mathrm{d}s' \tag{4.25}$$

and derived the exact solution

$$Z_{\mathrm{exact}}(\theta) = \frac{Z_0}{\left(1 + \frac{\tilde{Z}+Z_0}{\tilde{Z}t_0}\theta\right)^2}, \qquad Z_0 = \frac{1}{\frac{v_{\mathrm{in}}}{\hat{N}} - \frac{1}{\tilde{Z}}}. \tag{4.26}$$

Since the wavefront is traveling through one element within one time step, the impedances of the RM depending on the traveling time are expressed by

$$Z(\theta = q\Delta t) = Z_q, \quad q = 0 \ldots n-1. \tag{4.27}$$

The impedance results are depicted in Fig. 4.6. Both curves are continuously decreasing beginning from their maximum value at $\theta(0)$. Physically, it is comprehensible that the

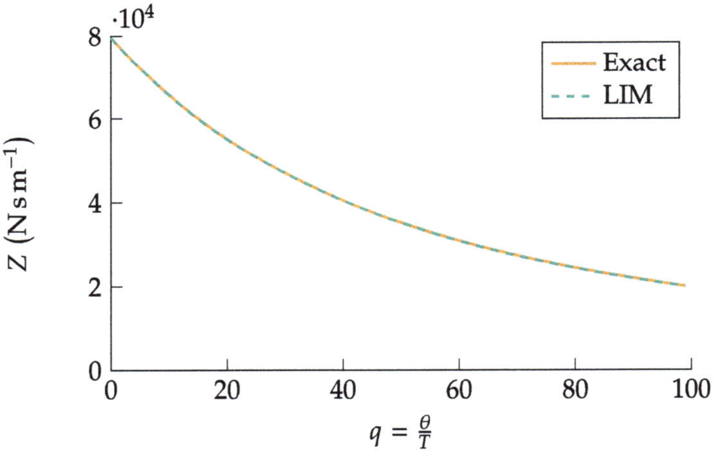

Figure 4.6: Comparison between the impedance function that is obtained with the RM and the exact solution.

curves decrease as the claimed impact force is decreasing exponentially as well. The impedance result determined with the RM matches very well the exact result. In fact, the maximum relative error is $1.10 \cdot 10^{-3}$ %. The remaining difference to the exact solution is due to the piecewise constant approximation of the rod's impedances. The more elements are chosen the closer the results of the RM approach the exact solution.

An extract of the results at designated time steps is presented in Tab. 4.3. The simulation starts with the determination of the first element of Rod 1 and finishes with the last element ($q = 100$). From $q = 10$ to until $q = 100$, every tenth result is plotted.

q	K_q	$\lvert \Delta N_q \rvert$ in kN	$\lvert K_q N_0 \rvert$ in kN	Z_q in $\mathrm{kN\,s\,m^{-1}}$
1	1.00	-	239	79.7
10	0.999	2.20	239	65.7
20	0.998	2.01	239	55.1
30	0.998	1.86	238	46.9
40	0.997	1.72	238	40.4
50	0.997	1.60	238	35.2
60	0.996	1.50	238	30.9
70	0.996	1.41	238	27.3
80	0.996	1.33	238	24.4
90	0.995	1.26	238	21.9
100	0.995	1.20	238	19.9

Table 4.3: Simulation steps of the example with exponentially decreasing impact force.

There is a clear tendency to be observed: All values, that is the transmission factor K_q, the force difference $\lvert \Delta N_q \rvert$, the force reserve $\lvert K_q N_0 \rvert$ and the impedances Z_q are continuously decreasing. In Sec. 4.1.3, it has been proven that K_q is continuously decreasing. Therefore, the results are consistent with the theory. At the beginning of the simulation, the change in impedance is greatest. That is the reason, why K_q is decreasing strongest at that time. As a result of the decreasing factor K_q, the force reserves have to fall as well because N_0 is constant. The exponentially declining impact force leads to continuously declining impedances. The calculated values reveal that the presented example did make very little use of the force reserves.

Example 2: Linearly Increasing Impact Force

The second example only differs from the first one by the impact force which is now linearly increasing

$$N_{\mathrm{I}}(t) = -\hat{N}\left(1 + 2\frac{t}{t_0}\right). \tag{4.28}$$

In Fig. 4.7, the related impedance result is plotted on a logarithmic scale. At $q = 78$, the simulation ends. Afterward, the calculated impedance value is negative which is

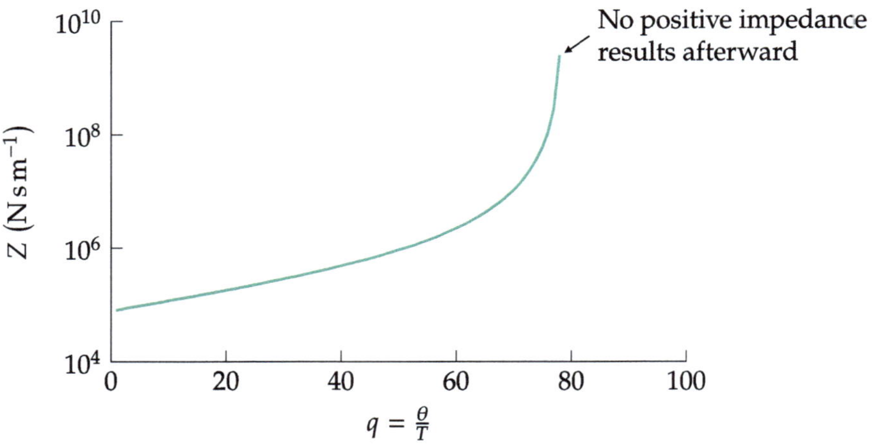

Figure 4.7: Impedance function of the example with linear increasing impact force.

not physical. Shortly before, the impedance values increase strongly. A more detailed overview of the results at designated simulation steps is presented in Tab. 4.4. The simulation starts with the first element of Rod 1 and finishes for $q = 78$. From $q = 10$ until $q = 70$, every tenth result is plotted. At the time of interest, from $q = 75$ on, every result is depicted. The absolute value of the force difference $|\Delta N_q|$ and the impedances Z_q are continuously growing whereas the absolute value of the force reserve $|K_q\, N_0|$ and the transmission factor K_q are continuously decreasing. The linear increasing impact

q	K_q	$\lvert\Delta N_q\rvert$ in kN	$\lvert K_q\, N_0\rvert$ in kN	Z_q in MN s m^{-1}
1	1.00	-	239	0.0797
10	0.996	4.90	238	0.120
20	0.992	5.19	237	0.183
30	0.987	5.76	236	0.291
40	0.980	6.78	234	0.496
50	0.970	8.67	232	0.961
60	0.951	12.7	227	2.43
70	0.893	26.0	213	13.0
75	0.765	56.3	183	106
76	0.693	73.5	166	276
77	0.556	106	133	2460
78	0.202	191	48.3	-

Table 4.4: Simulation steps of the example with linearly increasing impact force.

force leads to a continuous rise in the force difference and eventually to a continuous rise of the impedances. At the beginning, the increase is moderate, however, the longer the simulation takes the higher are the jumps in impedance. Especially, starting from $q = 70$, sharp increases occur. For $q = 78$, the existence condition is not met anymore. Therefore, no physical (positive) impedance can be determined.

One of the advantages of the presented approach is that the force condition immediately indicates if there will be a solution for the next time step. Moreover, it is possible to examine how long the given impact force can be generated. In this case, the maximum traveling time of the wavefront is $77\Delta t$.

4.2.2 Percussion Drilling

Percussion drilling has a long tradition and is widely used for drilling applications. The classical setup (see Fig. 4.8) consists of a pneumatically or hydraulically actuated piston which hits onto the shank adapter at hitting frequencies between 50 and 100 Hz. The shank adapter together with the coupling sleeve transfer the kinetic energy of the piston into the rod. The resulting waves propagate through the drill rod to the drill bit and finally fracture the ground due to high impact forces. The emerging cuttings are flushed up by a high pressured fluid that is pumped to the ground through hollow rods. The rotation of the drilling rod at rotation speeds between 60 and 200 rpm leads to continuously changing points of impact [61]. Because of the energy losses during the transmission of the shock waves, percussion drilling is used for drilling lengths no longer than 30 m [37].

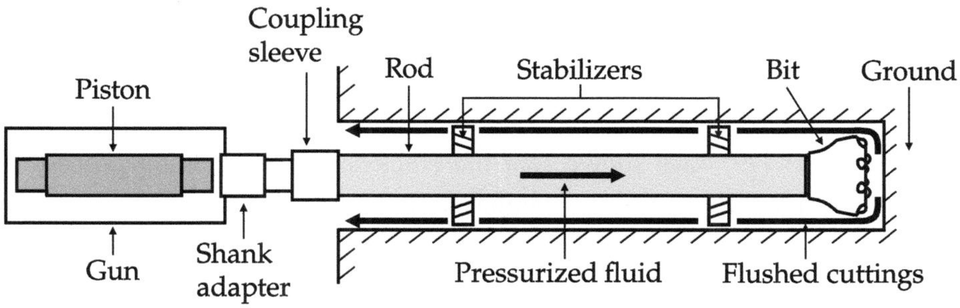

Figure 4.8: Sketch of the percussion drilling process, turned by 90°.

Problem Description and Modeling

The mechanics of percussion drilling has been studied for many years. However, the research has mainly focused on the direct problem. The modeling is either realized with a continuous model [9, 49], with an FEM model [13] or with lumped parameters [20, 42]. Comparisons between the results based on the 3D FEM and results based on the 1D theory show that the differences are very little so that the 1D theory is applicable [19, 48].

The considered inverse problem is depicted in Fig. 4.9. A piston of given length (0.5 m) and given initial velocity v_{in} is hitting onto the shank adapter. The shank adapter is modeled by a segment of constant impedance $\tilde{Z}_1$, the coupling sleeve by a segment with impedance $\tilde{Z}_2$ and the section with constant impedance $\tilde{Z}_3$ corresponds to the rod. The bit is composed of a constant impedance part $\tilde{Z}_4$ and a linear increase from $\tilde{Z}_4$ to $\tilde{Z}_5$ at the end. The bit-rock interaction during loading is modeled by a resistance force F with coefficient k_1. Its validity has been confirmed experimentally [17, 18, 39].

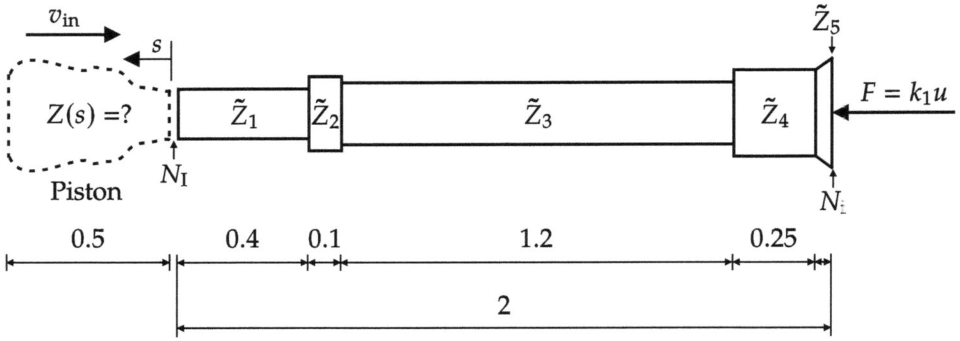

Figure 4.9: Example of a device for percussion drilling. The main parts are modeled as rod elements of different diameters. All dimensions are in m.

Lundberg and Collet [49] have shown that the efficiency of the drilling process is maximized if the incident force is of exponential shape

$$N_{\mathrm{i}}(t) = -\hat{N}\,\mathrm{e}^{Ct} \tag{4.29}$$

with an arbitrary initial force $\hat{N}$ and a constant C. The investigation motivates the realization of an exponential incident force by adjusting the impedance of the piston.

The simulation is run with the parameters listed in Tab. 4.5. The chosen coefficient k_1 and the constant C are characteristic for a hard rock [49]. The simulation starts with

Parameters		Number	Unit
$\tilde{Z}_1$	Impedance of the shank adapter	51.0	$\mathrm{kN\,s\,m^{-1}}$
$\tilde{Z}_2$	Impedance of the coupling sleeve	115	$\mathrm{kN\,s\,m^{-1}}$
$\tilde{Z}_3$	Impedance of the rod	79.7	$\mathrm{kN\,s\,m^{-1}}$
$\tilde{Z}_4$	Impedance of the constant bit segment	156	$\mathrm{kN\,s\,m^{-1}}$
$\tilde{Z}_5$	Impedance at the end of the bit	248	$\mathrm{kN\,s\,m^{-1}}$
$\hat{N}$	Initial incident force	187	kN
C	Constant	5.88	$\mathrm{(ms)^{-1}}$
k_1	Resistance force coefficient	300	$\mathrm{MN\,m^{-1}}$
$\tilde{n}$	Number of elements of the rod	400	—
n	Number of elements of the piston	100	—
$\tilde{v}_{\mathrm{in}}$	Initial velocity of the rod	0.00	$\mathrm{m\,s^{-1}}$
v_{in}	Initial velocity of the piston	9.00	$\mathrm{m\,s^{-1}}$

Table 4.5: Parameters used for the percussion drilling example.

the transformation of the incident force N_{i} to the corresponding impact force N_{I}. This enables applying the RM to calculate the piston's impedance function subsequently.

Transformation of the Incident Force to the Impact Force

In Fig. 4.10a, the incident force described in Eq. (4.29) and the transformed impact force are depicted. For better comparability of the results, the incident force is time-shifted by the dead time t_{dt} which is the travel time from the impact surface to the bit-rock interaction.

The curve of the impact force is above the curve of the incident force. The reason is that the forces applied to the impact surface are amplified while traveling to the bit-rock interaction. The impact force curve exhibits three major jumps whose origins are depicted in Fig. 4.10b. The first jump occurs when the double reflected waves at the coupling sleeve reach the bit-rock interface. The second jump is significantly larger. It originates from the waves that are first reflected at the bit-rock interface, subsequently reflected at the transition between bit and rod and finally travel back to the bit-rock transition. These waves are compressive which is the reason why the magnitude of the impact force decreases afterward. The last jump is due to the double reflected waves between the impact surface and the coupling sleeve.

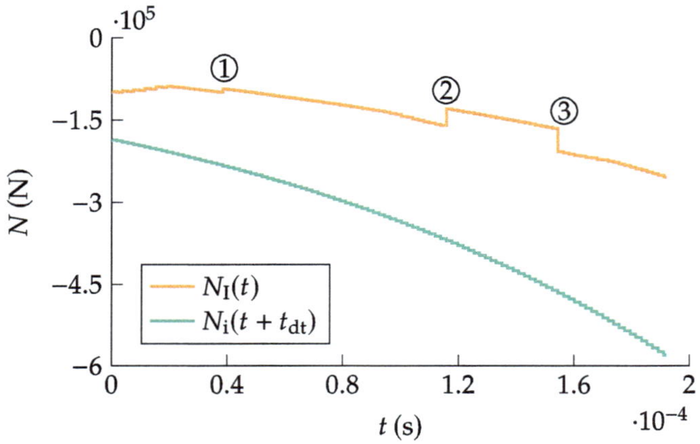

(a) Transformation from the incident force N_i to the impact force N_I. The incident force is time-shifted by the dead time t_{dt} which is the travel time of the propagating waves from the impact surface to the bit-rock transition.

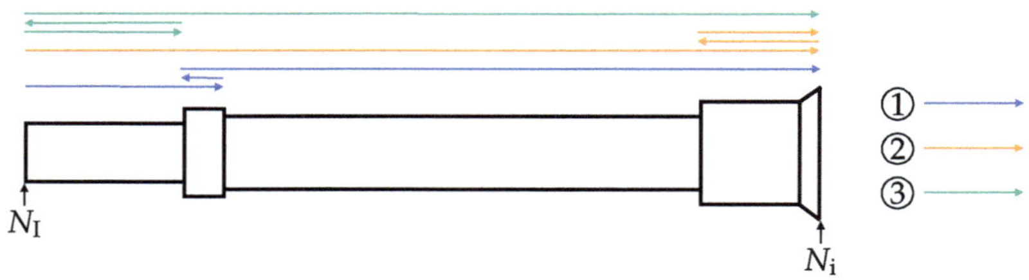

(b) Explanation of the most significant points of N_I.

Figure 4.10: Transformation of the incident force to the impact force.

Application of the Recursive Method

As the incident force has now been transformed to the corresponding impact force, the RM is applied. In Fig. 4.11, the impedance over the distance s (starting for $s = 0$ at the hitting surface) of the optimized piston is shown. As expected, the impedance is increasing as the absolute value of the impact force is increasing as well. The three jumps also emerge in the impedance curve. This is sensible since the jumps in piston impedance lead to the desired jumps in impact force.

Using the RM, the results are determined exactly. This is confirmed by a comparison of the desired incident force with the incident force obtained by running the simulation with the determined impedance function of the piston. The maximum percentage error is $1.13 \cdot 10^{-12}$ % which is in the order of machine precision. Moreover, it is possible to determine how long a solution of the inverse problem exists. For the chosen piston length, a solution exists. However, the longer the piston is, the higher are the

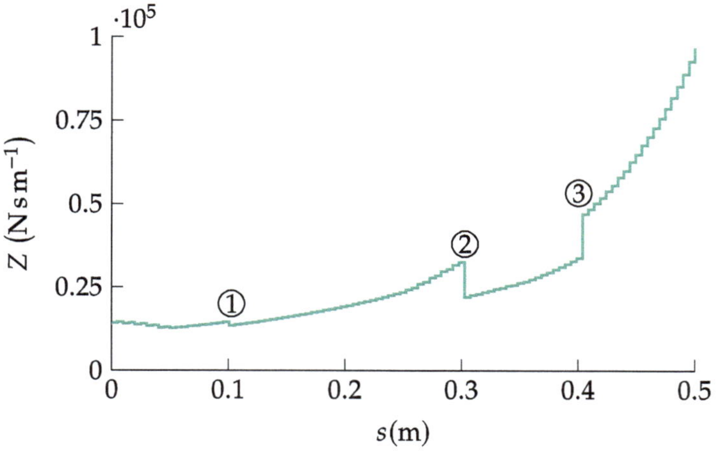

Figure 4.11: Calculated impedance function of the piston.

impedances at the end of the piston. At some point, no solution exists anymore. The huge difference between maximum and minimum impedance restricts the practical applicability. Besides, it is only possible to generate the desired incident force twice the travel time through the piston. Afterward, no elements are left to shape the curve. However, the percussion drilling example has shown that the RM is also applicable if the desired force is prescribed at a position other than the impact surface. Besides, the impedance function of the rod may vary and linear boundary conditions may be considered.

5 Experimental Investigation

In this chapter, the 1D model is experimentally validated. After presenting the experimental setup with its components, preinvestigations are considered. Several aspects have to be examined in advance to guarantee reliable results. The most important factor is the repeatability of the experimental results as non-repeatable results have limited significance. The strains are measured by strain gauges (SGs) that are attached to different circumferential and axial positions of the rod. It is assessed if bending or the influence of the rod support results in different SG measurements. Subsequently, it is evaluated if the results can be scaled according to the impact velocity. The preinvestigations finish with the determination whether the piston bearings influence the results.

After the preinvestigations, two different experiments, which only differ from the cross sectional area variation of the piston, are conducted. Finally, the measurements are evaluated and compared with simulation results obtained by applying the wave propagation method.

5.1 Experimental Setup

The test rig consists of a gun that pneumatically accelerates a piston which finally hits the rod (see Fig. 5.1). The impact velocity of the piston is measured by a photoelectronic fork sensor called *SpeedGate* of the manufacturer *Frederikson*. The sensor's two vertical laser rays at known axial distance are cut by the front face of the piston at different times. As both the time difference of these cuts and the related axial distance is known, the impact velocity can be determined. After the hit, force waves propagate into piston and rod. At three different axial positions, SGs, that are only able to detect longitudinal elongation, are attached to the rod. Their gauge length (0.2 mm) is very short which ensures good spatial resolution.

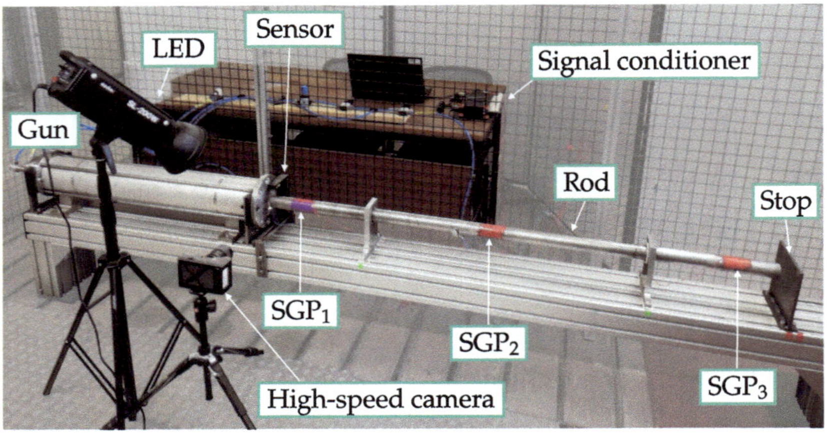

(a) Experimental setup.

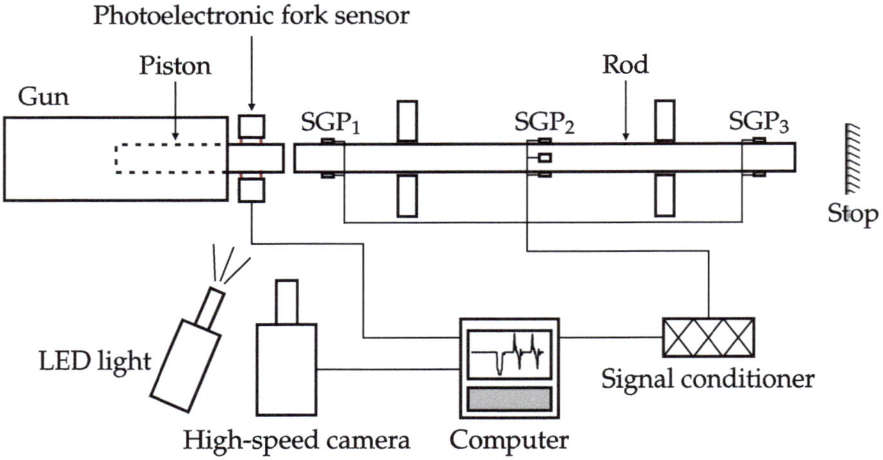

(b) Schematic sketch of the experimental setup.

Figure 5.1: (a) Experimental setup and (b) its schematic sketch.

Each SG is part of a quarter bridge (see Fig. 5.2) that transforms the mechanical strain into electrical voltage. The transformed electrical signal is amplified and filtered by an analog anti-aliasing filter (Bessel, 5th order) with corner frequency 100 kHz. Subsequently, the signal is digitalized and filtered by a digital IRR filter. The filter properties of the analog filter are predefined whereas the properties of the digital filter are set according to the application. For the given example, a conflict of interest exists. On the one hand, the corner frequency of the digital filter (Bessel filter 4th order) has to be very high as the

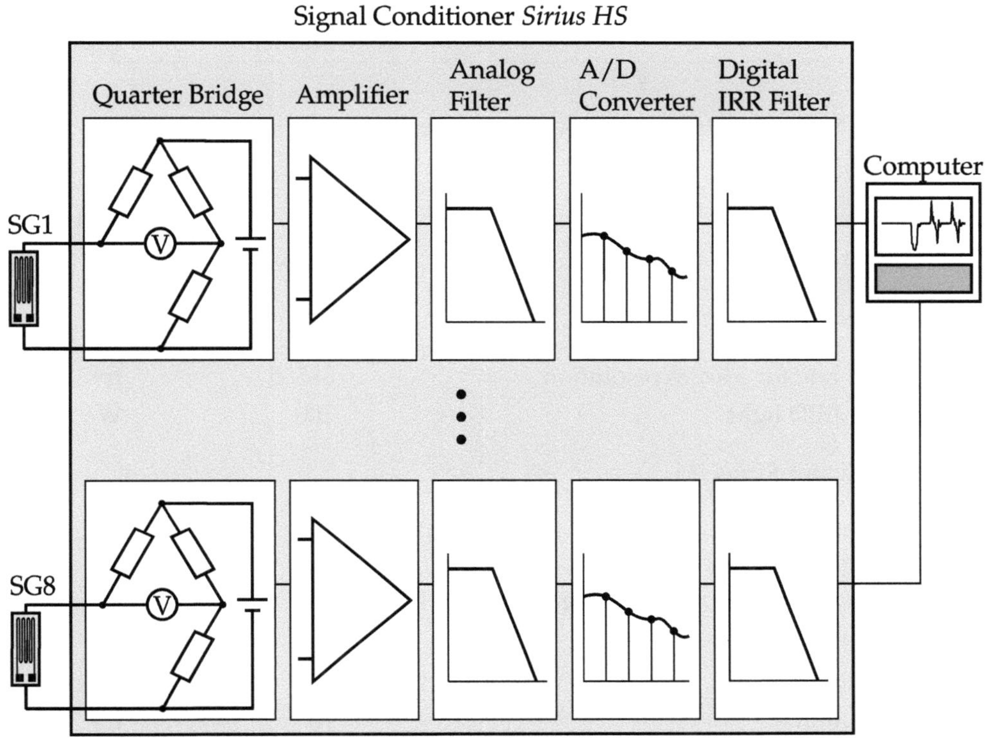

Figure 5.2: Measuring chain from the SG to the representation at the computer.

transient strain signal includes high frequencies. On the other hand, the higher the corner frequency that is chosen the more noise is included in the signal. Based on these conflicting goals, the corner frequency 40 kHz is set. The signal processing is conducted with the signal conditioner *Sirius HS* from *ZSE*. Finally, the measurements are evaluated by a computer.

The high-speed camera records the hit at its maximum frame rate (38565 fps) which enables detecting at what time piston and rod separate. Moreover, the videos provide very good insight during the impact. As the ambient light is not sufficient for these frame rates, external LED-lights illuminate the impact zone. The parameters of the SGs, the high speed camera and the signal conditioner are summarized in Tab. 5.1.

The rod can move freely at its right end. The stop only prevents the rod from crashing into the fence after measuring. A more detailed examination of the piston and rod and of the pressure chamber is provided in the next subsections.

Parameters	Number	Unit
Strain gauge *KFGS-02-120-C1-11L3M2R*		
Gauge factor	2.23 ± 1 %	–
Gauge length	0.2	mm
Gauge resistance	119.6 ± 0.4 %	Ω
High speed camera *Chronos 1.4*		
Frame rate for 336*x*96 resolution	38565	fps
Power LED light	200	W
Signal conditioner *Sirius HS*		
Corner frequency of analog anti-aliasing filter (Bessel, 5th order)	100	kHz
Corner frequency of digital IIR filter (Bessel, 4th order)	40	kHz
Sampling rate	1	MHz
Resolution	16	Bit

Table 5.1: Parameters of the SGs, the high speed camera and the signal conditioner.

Piston and Rod

For the experimental results, one rod and two different pistons are applied. The dimensions of the pistons are shown in Fig. 5.3 and Fig. 5.4, respectively. Piston 1 is symmetrical with two small diameter parts (40 mm) at their ends and one large diameter part (50 mm) in the middle. Piston 2 consists of many cross section area parts, including several steps and one cone.

The dimensions of the rod and the axial strain gauge positions (SGPs) are depicted in Fig. 5.5. The rod's diameter is constant (50 mm) with a total length of 1800 mm. The axial SGP that is closest to the impact zone is named SGP_1, the one in the middle of the rod SGP_2 and the one to the very end of the rod is named SGP_3. The SGs are attached symmetrically meaning that the distance of SGP_1 from the left end equals the distance of SGP_3 from the right end (each 110 mm). Furthermore, SGP_2 is exactly in the middle of the rod. The SGs are mounted at different circumferential positions. They are either

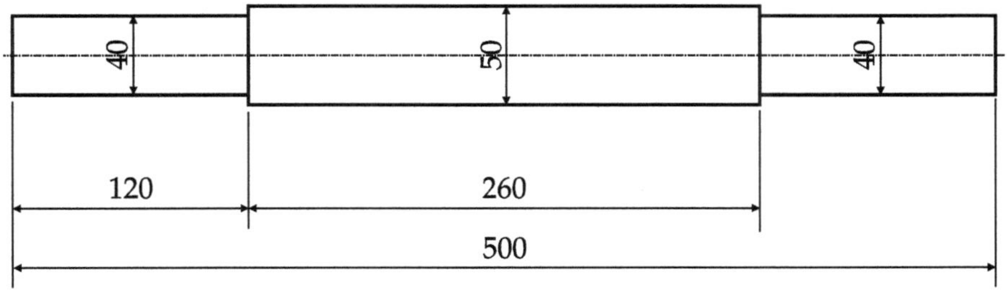

Figure 5.3: Dimensions of Piston 1 in mm.

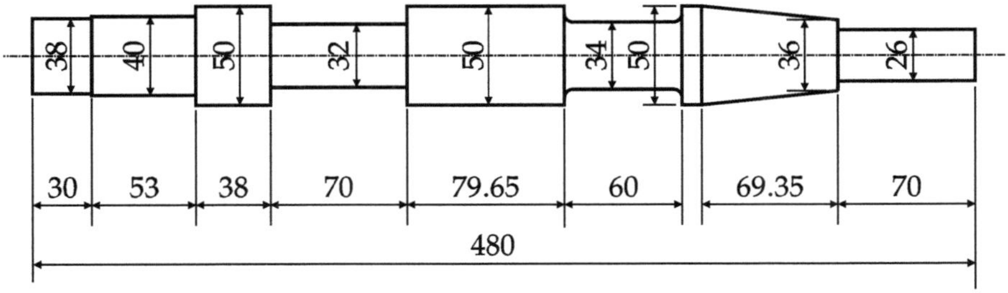

Figure 5.4: Dimensions of Piston 2 in mm.

mounted on the top (0°), right (90°), bottom (180°) or left (270°) side of the rod (see Fig. 5.5). Overall, eight SGs are attached to the rod. Two SGs are attached to SGP_1 (top, bottom), four SGs are attached to SGP_2 (top, right, bottom, left) and two SGs are attached to SGP_3 (top, bottom). The SGs are named by SGi,x, where i refers to the SGP and x to the angle. For example, SG1,90° refers to the right SG at SGP_1.

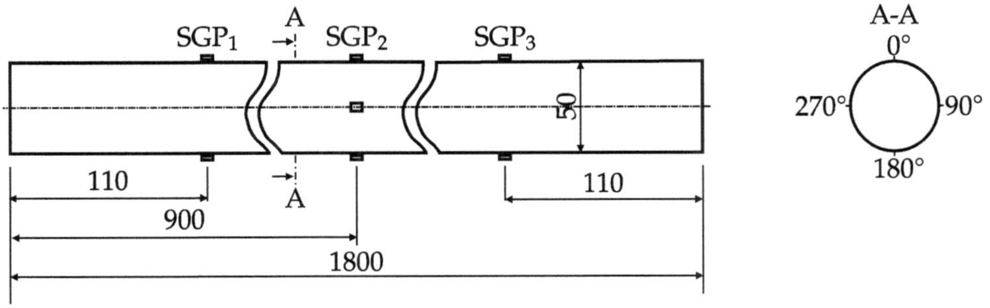

Figure 5.5: Dimensions of the rod and the axial position of the SGs in mm (left) and a sketch of the circumferential positions of the SGs (right).

The pistons and the rod are made of steel with the parameters listed in Tab. 5.2.

Parameters		Number	Unit
E	Young's modulus	210	GPa
ρ	Density	$7.84 \cdot 10^3$	kg/m^3

Table 5.2: Parameters for the pistons and the rod.

Pressure Chamber

The acceleration of the piston is carried out pneumatically. Between the outer and inner diameter of the gun, a pressure chamber is filled with pressurized air (see Fig. 5.6). The chamber is kept closed by a bushing which is actuated pneumatically until the pressure in the chamber reaches the desired air pressure. Once the pneumatic actuator pulls the

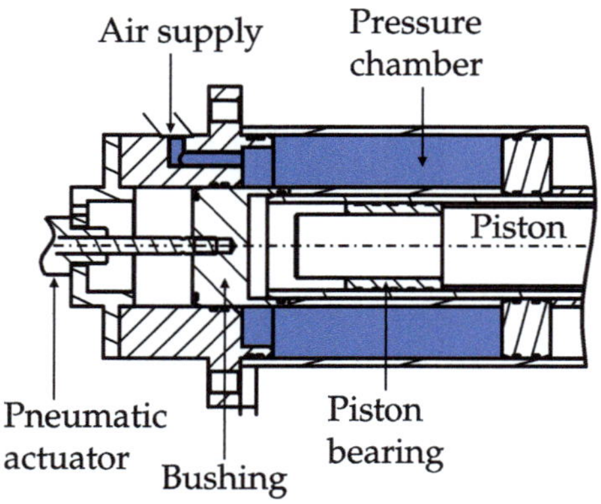

(a) Closed.

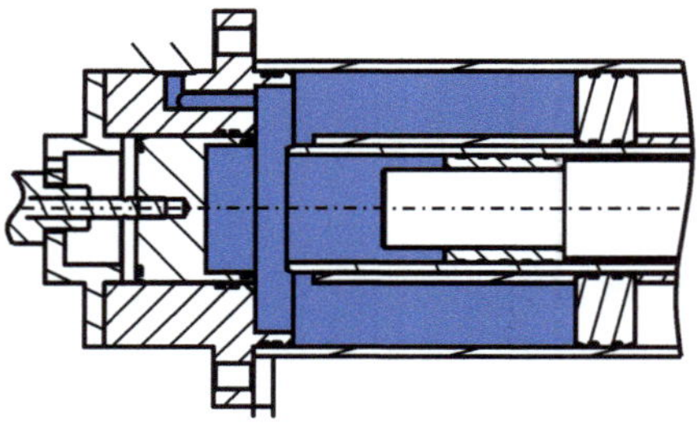

(b) Open.

Figure 5.6: Pressure chamber of the gun in (a) closed and (b) open state.

bushing to the left, the pressurized air is flowing behind the left end of the piston which leads to its acceleration. The experimental setup is a single hit device meaning that the piston is not automatically moving back and forth. Moreover, as the pressure in the pressure chamber is controlled, it is not possible to control the piston's impact velocity directly.

5.2 Experimental Procedure

Two different setups are investigated. The first setup (see Fig. 5.7) only differs from the second setup (see Fig. 5.8) in the shape of the impacting piston. The rod, which initially is at rest, can move freely at its right end. The stop does not influence the results as only the transient behavior of the waves immediately after the impact is analyzed. The initial velocity of the pistons may be changed according to the investigated example.

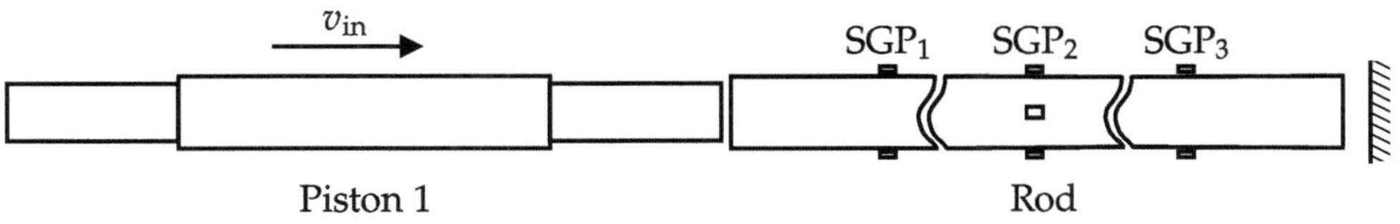

Figure 5.7: Setup 1.

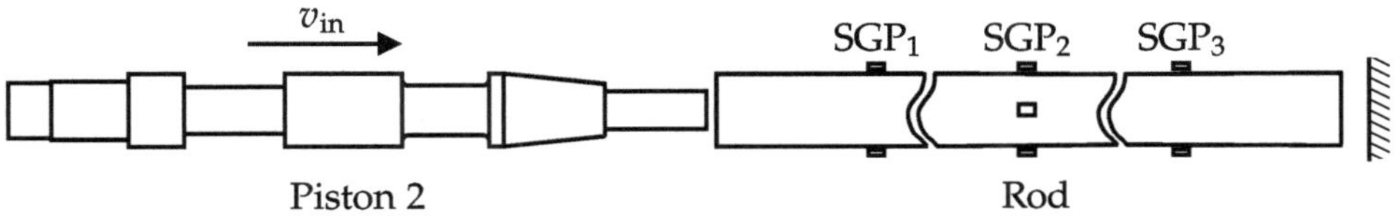

Figure 5.8: Setup 2.

For the evaluation of the results, the normal forces are calculated from the measurements at the SGs. The normal force $N_{i,x}$ refers to the force at SGi,x. For example, $N_{2,0^\circ}$ is the normal force at SG2,0°. For most evaluations, only the normal force N_k, which compensates bending, is denoted. It is determined with

$$N_k = \frac{1}{2}\left(N_{k,0^\circ} + N_{k,180^\circ}\right), \; k = 1\ldots3. \tag{5.1}$$

The subscript number of the forces refers to the corresponding SGP.

The SGs are attached to the top and bottom at each SGP to both compensate the normal forces for bending and to determine the maximum bending stress

$$\sigma_{k,\text{bend}} = \frac{1}{2\,A_{\text{Rod}}}\left(N_{k,0^\circ} - N_{k,180^\circ}\right), \; k = 1\ldots3, \tag{5.2}$$

with the cross section area A_{Rod} of the rod. By using two diametrically opposite strain gauges in a half bridge, it would not be possible to determine the maximum bending stresses which is why a quarter bridge has been used.

In Tab. 5.3, an overview of the conducted experiments is given. The preinvestigations are all carried out with the first setup. The *repeatability* experiment is conducted three times. Since it is not possible to directly control the impact velocity, little differences in

Name	Setup	v_{in} ($m\,s^{-1}$)	Width Bearings (mm)	Evaluated Values
Preinvestigations				
Repeatablity	1	4.57	60	N_1
		4.58	60	N_1
		4.60	60	N_1
Circumfertential SGP	1	4.60	60	$N_{2,0°}, N_{2,90°}, N_{2,180°}, N_{2,270°}, \sigma_2, \sigma_{2,bend}$
Impact Velocity	1	3.47	60	$N_1/\lvert \min(N_1)\rvert$
		4.60	60	$N_1/\lvert \min(N_1)\rvert$
		5.60	60	$N_1/\lvert \min(N_1)\rvert$
Piston Bearings	1	4.60	6	N_1
		4.61	15	N_1
		4.62	60	N_1
Main Investigations				
Standard Piston	1	4.60	60	N_1, N_2, N_3
Complex Piston	2	4.58	60	N_1, N_2, N_3

Table 5.3: Parameters of the conducted experiments.

velocity occur. However, the deviations are negligible. The corresponding results for the normal force at SGP_1 are compared. The *circumferential SGP* experiments aim at both determining the influence of the bending stress and the influence of the rod bearings. Thus, the four normal forces at SGP_2 are compared and the ratio $\max\left|\sigma_{2,\text{bend}}\right|/\max\left|\sigma_2\right|$ with $\sigma_2 = N_2/A_{\text{Rod}}$ is determined. The *impact velocity* experiments differ in the impact velocity which range from $3.47\,m\,s^{-1}$ to $5.60\,m\,s^{-1}$. It is assessed if the results are scalable by comparing the normalized normal forces at SGP_1. The *piston bearings* experiments give information on whether the piston bearings have to be considered in the model by comparing the results for N_1.

In the main investigations section, the normal forces N_1, N_2 and N_3 are analyzed and interpreted both for Setup 1 and Setup 2. At the end, the experimental results for N_1 are compared with the corresponding simulation results which are obtained by applying the wave propagation method.

5.3 Preinvestigations

The focus of the preinvestigations is on determining possible influence factors on the measurements. Therefore, the interpretation of the results is qualitatively and comparative. A detailed interpretation of the results will be conducted in Sec. 5.4.

5.3.1 Repeatability

In Fig. 5.9, the experimental results at SG1,0° for three impacts are depicted. The results

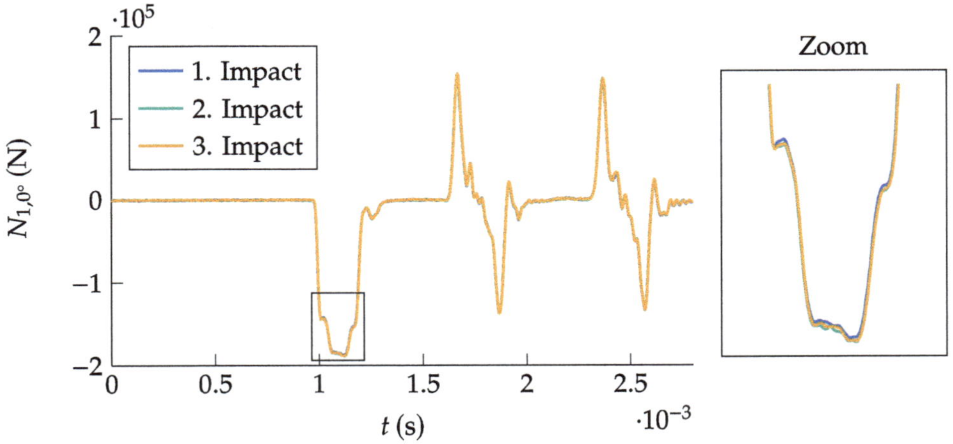

Figure 5.9: Examination of the repeatability of the mesaurement results. The initial velocities of the piston are in the range between 4.57 m s^{-1} and 4.60 m s^{-1}.

match very well without any obvious deviations. Very similar results are achieved when the measurements are analyzed at other axial and other circumferential positions. Therefore, it is concluded that the results are repeatable.

5.3.2 Influence of Circumferential Strain Gauge Position

The SGs are placed at different circumferential positions. In theory, a direct central impact only causes longitudinal waves to propagate into the rod. However, optimal conditions are never met. Furthermore, the rod is supported at two positions which may influence the results due to friction. This examination aims to assess if the rod bearings have to be considered in modeling and if the impact is direct central. Therefore, the results at the second SG for four different circumferential positions are shown in Fig. 5.10.

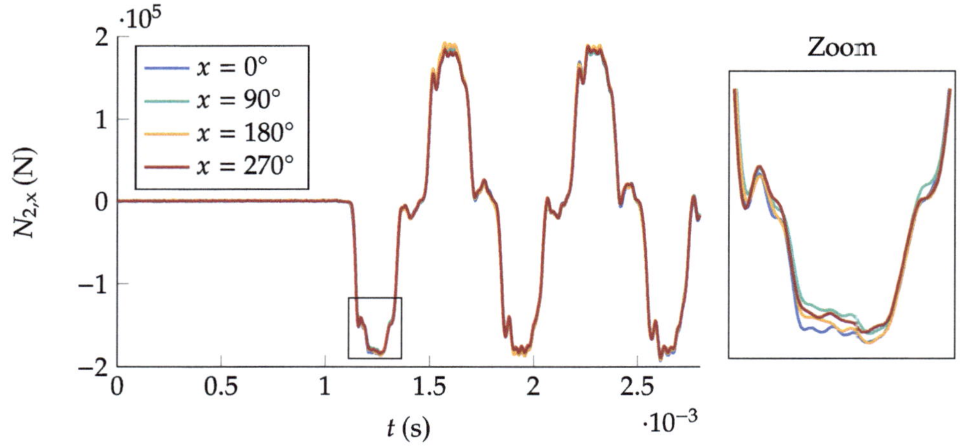

Figure 5.10: Influence of the circumferential SGP on the experimental results. The initial velocity of the piston is 4.60 m s^{-1}.

The qualitative behavior of the results is almost the same for all angles. Quantitatively, the deviations are very small which can be regarded in the zoom window. Therefore, it is concluded that the influence of the rod bearing is negligible and that the impact is close to direct central. This assumption is confirmed by the calculated ratio

$$e = \frac{\max\left|\sigma_{2,\text{bend}}\right|}{\max\left|\sigma_2\right|}100 \tag{5.3}$$

between maximum bending stress and normal stress which is 2.74 %.

5.3.3 Influence of Impact Velocity

The initial velocity of the impacting piston varies for different applications. As the presented test rig is also restricted in terms of its maximum piston velocity, the question arises if the results are scalable. In Fig. 5.11, the measurement results obtained with three different initial velocities are depicted. The results are normalized by the absolute value of the respective global minimum. The deviations between the normalized results are very low which is highlighted in the zoom box. This allows the conclusion that within the presented velocity range from 3.47 m s^{-1} until 5.60 m s^{-1} the results are scalable. However, it is expected that the results are also scalable for a wider range but this needs finally to be proven by other experiments.

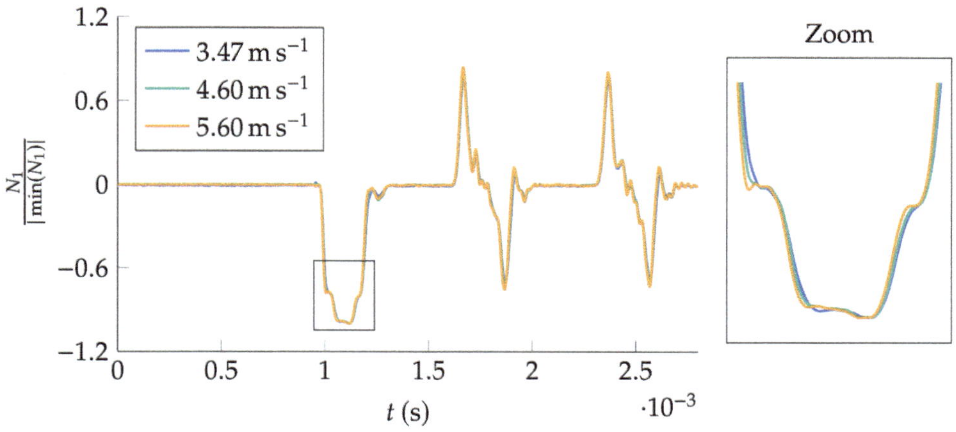

Figure 5.11: Influence of the initial velocity of the piston on the experimental results.

5.3.4 Influence of Piston Bearings

The pistons are mounted with bearings made of Polytetrafluorethylen as can be seen in Fig. 5.12. Since it is not possible to accelerate the piston without bearings with the presented test rig configuration, it has to be clarified if the piston bearings influence the measurements. If the bearings influenced the results, there would be a difference in the measurements depending on the bearing width. However, as can be seen in Fig. 5.12,

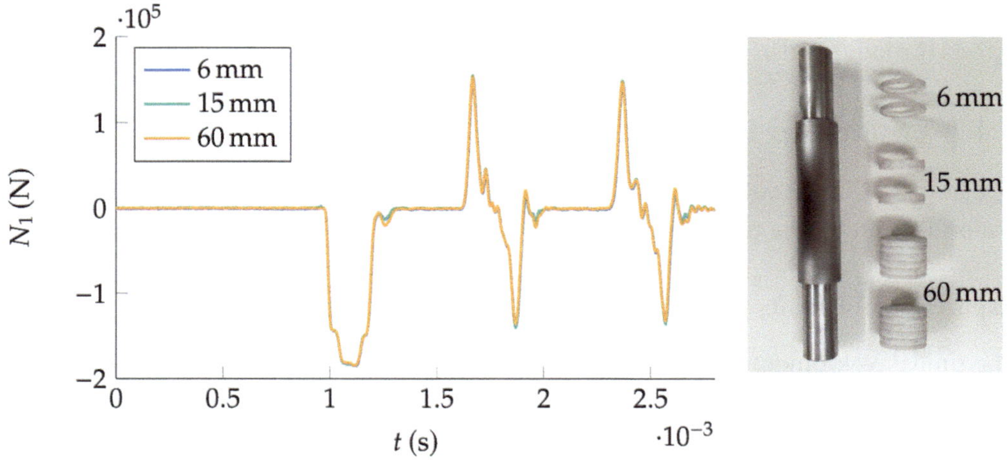

(a) Measurement results for different bearing widths.

(b) Piston 1 and its bearings.

Figure 5.12: Influence of the bearings of the piston on the experimental results. The initial velocities of the piston are in the range between 4.60 m s^{-1} and 4.62 m s^{-1}.

there is no significant difference apparent which is why the bearings are not considered.

In summary, the preinvestigations reveal that the experimental results are

- repeatable,
- negligibly independent of the circumferential position,
- scalable with respect to impact velocity,
- and not influenced by the piston bearings.

5.4 Main Investigations

The goal of the main investigations is to validate the 1D model presented in Chap. 3. Therefore, two experiments are conducted, analyzed and finally compared with the related simulation results.

The first experiment is conducted to assess if the experimental results are reliable. As the geometry of Piston 1 varies only very little, it is possible to compare the measurements with analytical results.

The second experiment is conducted with a complex piston geometry (Piston 2) with many changes of the cross section area in axial direction. Piston 2 is applied to evaluate if the 1D model is also valid for complex piston shapes.

5.4.1 Standard Impacting Rod

The first experiment coincides with the setup used for the preinvestigations (see Fig. 5.7). Piston 1 is hitting the rod at initial velocity $4.60\,\mathrm{m\,s^{-1}}$.

Experimental Results

The time development of the forces, that are related to the corresponding SGs at three different SGPs, is illustrated in Fig. 5.13. The numbers on the graphs belong to selected time points. The qualitative behavior at these time points is explained with Fig. 5.14. At t_1, the first deflection is observed at the first SGP. After an abrupt drop of the force, the force remains almost constant for a short period. Subsequently, another step to the maximum compressive force follows. Apart from small oscillations after t_3, the wave shape between t_1 and t_3 is symmetrical. Just before t_1, the piston hits the left end of

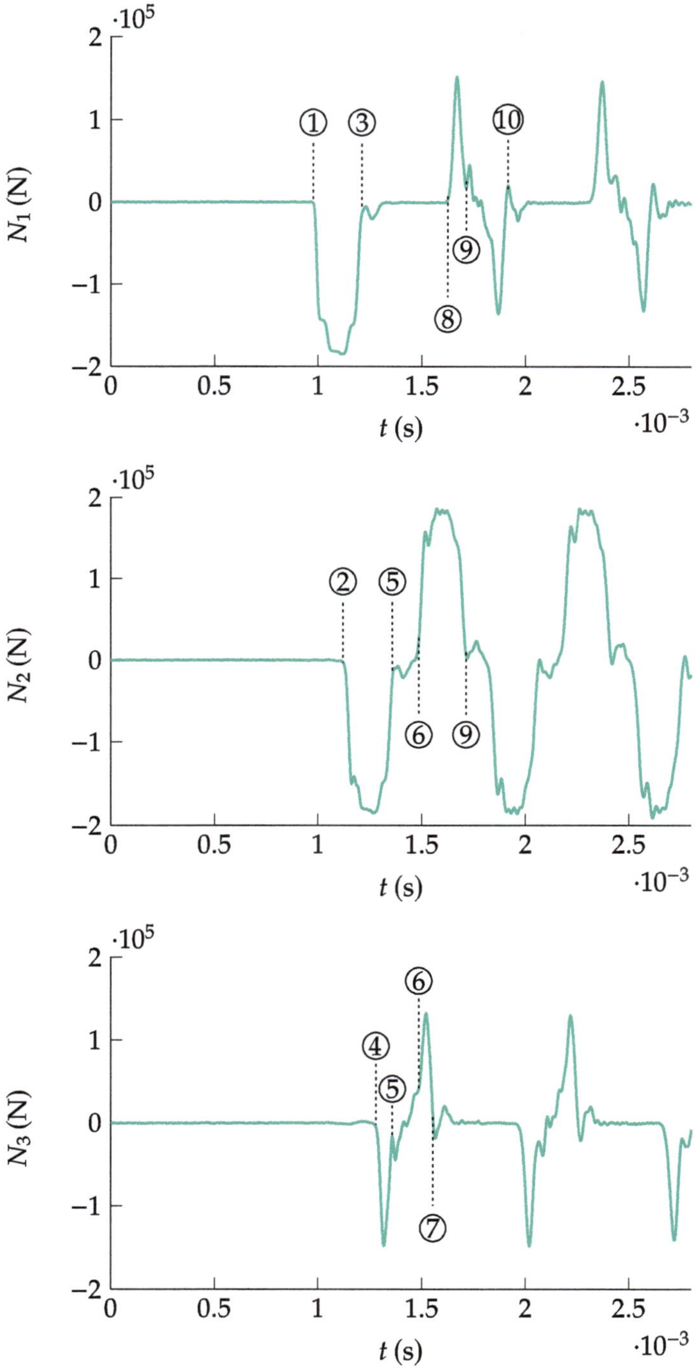

Figure 5.13: Forces versus time (first experiment) gained at the SGPs. The initial velocity of the piston is 4.60 m s^{-1}.

the rod. The time from $t = 0$ until t_1 arises due to the pretrigger settings and shall not be confused with the traveling time of the waves from the hitting surface to the first SGP. During the time from t_3 to t_8, the force is close to zero. Subsequently, a point symmetrical section with a tensile force peak at the beginning and a compressive force peak at the end (between t_8 and t_{10}) is observed.

The incident wave shape at the second SGP from t_2 to t_5 is very similar but time-delayed with respect to the wave shape at the first SGP between t_1 and t_3. Furthermore, the shape of the subsequent part ($t_6 - t_9$) resembles the incident wave with the opposite sign. The sequence from t_2 to t_9 also repeats shortly after.

The results at the third SGP lastly drop (starting from t_4). The following wave shape between t_4 and t_7 resembles the wave shape of the first SGP between t_8 and t_{10}. However, this time the wave shape begins with a compressive force peak and ends with a tensile peak which is reverse at the first SGP. Again, the wave shape between t_4 and t_7 repeats some time later.

For understanding the qualitative behavior of the force waves, it is advantageous to study the hit of two rods of the same impedance. In Fig. 5.14, the forces at any piston and rod position are depicted for selected time points which refer to the time points marked in Fig. 5.13. The piston is located on the left side whereas the rod with its three SGPs is placed on the right side. The force wave propagation direction is indicated with small arrows and the colors of the waves state if a wave is compressional (green) or tensile (orange).

After the hit, compressional force waves travel both into piston and rod. At t_1, the wavefront reaches the first SGP and leads to a jump. It is impossible to perfectly represent a jump in the SG measurements due to four reasons. Firstly, the SGs always measure the average elongation along their gauge length. Secondly, unlimited sample rates cannot be chosen. Thirdly, the signal is filtered with a low pass filter. And most importantly, due to 3D-effects, there is always a finite rise in reality. Moreover, the filter causes oscillations after jumps. This has to be considered when interpreting the results.

At t_2, the wavefront of the rod reaches the second SGP leading to the same jump as at the first SGP. Meanwhile, the wavefront of the piston has traveled to the left end where the wave is reflected as a tensile force. The compressive and tensile force waves are equal in amount which is why the resulting force is zero. The tensile force wave continues

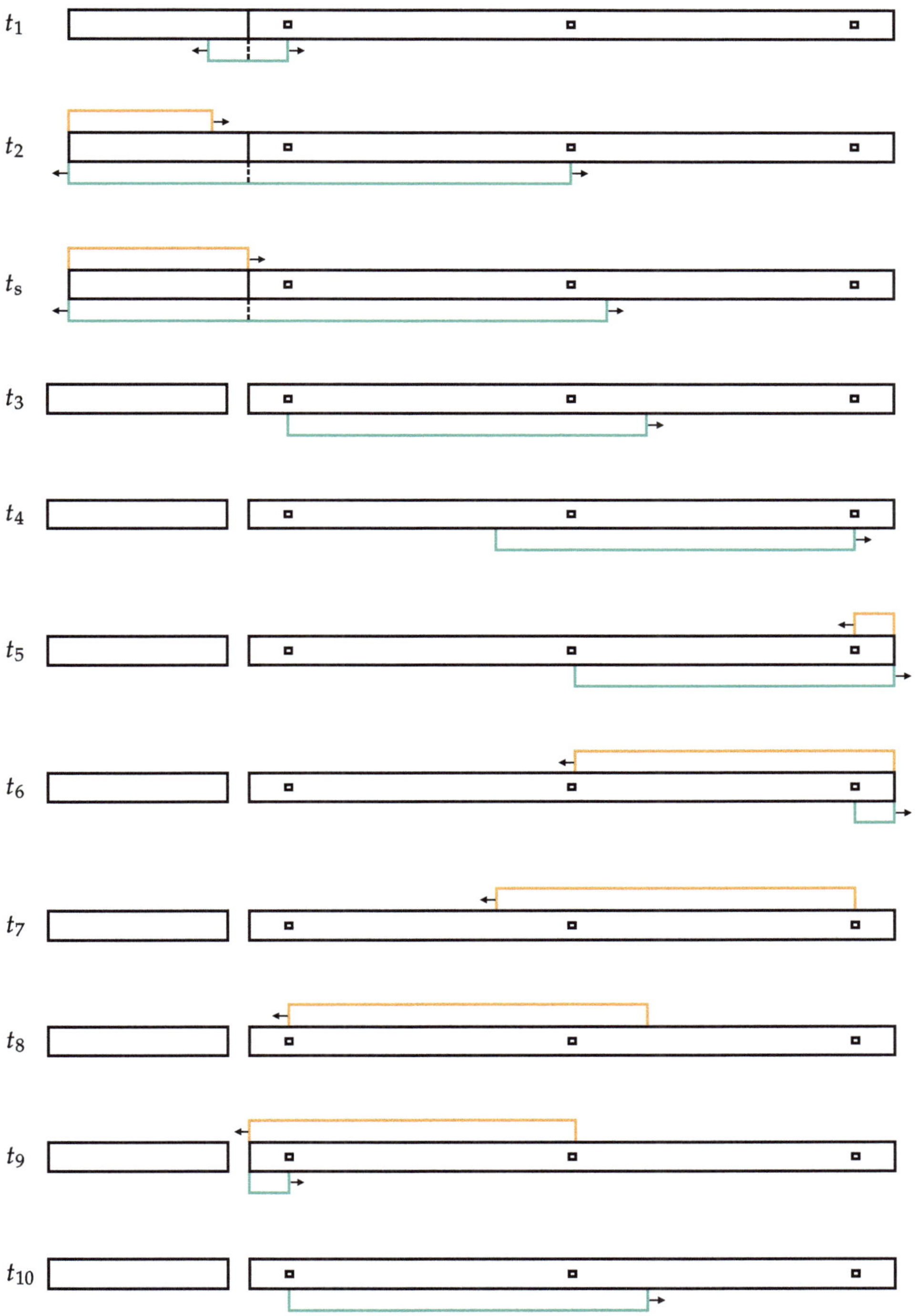

Figure 5.14: Schematic sketch of the wave propagation after a hit of two rods with same impedance.

traveling along the piston until it reaches the transition to the rod. This is the time t_s when piston and rod separate as no tensile force may be transmitted at the transition. Therefore, a rectangular wave shape that extends to double the length of the piston, propagates along the rod. The development of the forces in the piston is not pursued any longer.

At t_3, the compression wave passes the first SGP leading to zero force again. The time period t^* between the arrival of the wave front at the first SGP until the wave has completely passed the first SGP is determined with

$$t^* = \frac{2\ell_{\text{Pis}}}{c}. \tag{5.4}$$

For the given example, t^* equals $1.93 \cdot 10^{-3}$ s. This is in very good accordance with the time span between t_3 and t_1.

As the rod's impedance is constant, the wave shape does not change while traveling along the rod. Therefore, the wavefront, that arrives at the third SGP at time t_4, has the same shape when arriving at the other SGPs. At SGP2, no reflecting waves interfere which is why the shape of the normal force at the second SGP (between t_2 and t_5) is the time-delayed shape of the force at the first SGP (between t_1 and t_3).

At t_5, two events happen almost simultaneously. The incident compressive wave passes the second SGP and the reflected tensile force wavefront reaches the third SGP leading at both SGPs to zero resulting force.

When the compressive wave passes the last SGP at t_6, only the tensile force is acting on the SGs which is the reason why the force is jumping again. At the same time, the tensile wavefront reaches the second SGP which causes a sharp rise in tensile force.

When the reflected part of the incident wave completely passes the third SGP (time t_7), the force drops to zero. From time t_8 to t_{10}, the same procedure as for the third SGP takes place for the first SGP with the only difference that the incident wave is tensile initially.

At t_{10}, the state equals the state at t_3. Thus, the period between t_3 and t_{10} is repeated all over again until the oscillations are damped by, e.g., material damping. However, in the considered period, the influence of material damping is negligible. The period time complies with the time it takes to travel twice along the rod.

Comparison with Numerical Results

The measurements are compared with the simulation results which are generated by applying the wave propagation method. The parameters, the initial conditions and the setup of the simulation are set identically to the experiment. Moreover, the results are filtered by the same filter (Bessel 4th order) as the measurements are filtered. The

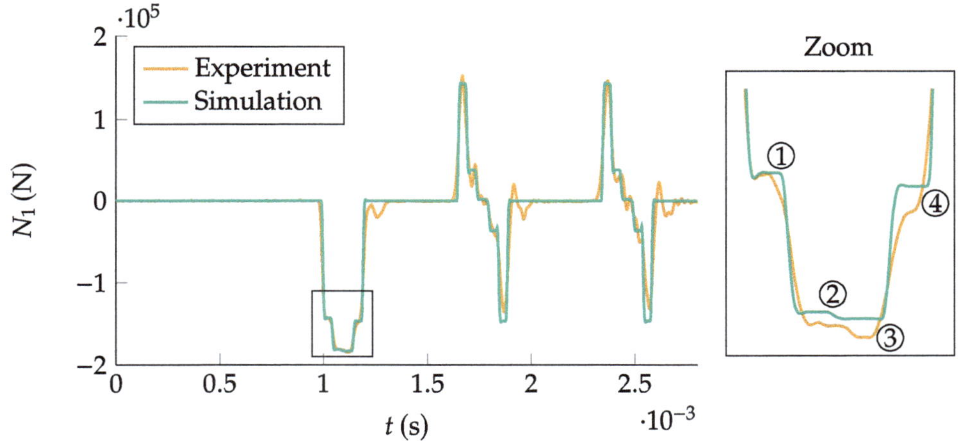

(a) Comparison between experimental and simulation results.The initial velocity of the piston is 4.60 m s^{-1}.

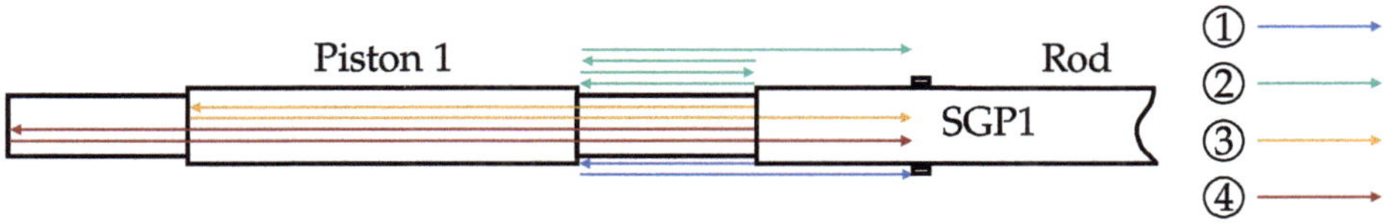

(b) Explanation of the development of the steps in the zoom box.

Figure 5.15: Evaluation of the first experiment.

resulting graph is plotted in Fig. 5.15a. A part of the first incident wave is highlighted in the zoom box. The deviations between the results are very little. The largest deviations are detected after jumps when the experimental solution still oscillates. Both the qualitative and the quantitative behavior of the curves matches very well as can be seen in detail in the zoom box. After the arrival of the incident compressive force, four main deflections are observed. Its sources are schematically visualized in Fig. 5.15b.

The first deflection originates from the wave that incidents in the piston, reflects at its first step and finally transmits to SGP_1. The second deflection is considerably smaller which is due to more reflections included in the development. The waves responsible for the first and second step are identical until the wavefront reaches the transition between piston and rod. While the first wave is transmitted to the SGs, the second wave is reflected back and forth twice between transition and the first step of the piston before it arrives at the SGs. The third wave is transmitted to the second step of the piston where it is reflected and transmitted directly to the SGs. The fourth wave is similar to the third one with the difference that the fourth wave is reflected at the free end of the piston. It is important to emphasize that the presented waves are only the ones that contribute most to the final shape of the force wave. The other waves are reflected very often so that they are very low in amplitude and negligible for the qualitative understanding of the final wave shape.

In summary, the accordance of the simulation with the experimental results is very good. Even small deflections are depicted by both approaches. Similar results are achieved at the remaining two SGPs. Therefore, the 1D model is validated for small variations in the piston geometry. In the next subsection, the question is answered if the statement is also valid for major variations of the geometry.

5.4.2 Complex Impacting Rod

The second experiment is similar to the first experiment but with a different piston hitting the rod (see Fig. 5.8). Piston 2 has several cross sectional area variations, including a conical increase at the beginning.

Experimental Results

In Fig. 5.16, the experimental results of the second experiment at three SGPs are presented. Since only the piston geometry varies compared to the first experiment, the qualitative behavior of the results is similar. An incident wave starting at $t \approx 1$ ms at the first SGP is traveling along the rod maintaining its shape while passing the second SGP. Over time, the incident wave shape continuously alternates between compressive and tensile force at SGP_2. At the third SGP, only the very beginning of the incident wave is visible as the measurement is superimposed by reflecting waves at the rod's free end a little while later. After two peaks (compressive and tensile), a section with forces close to zero follows before the signal is repeated. Very similar results occur at the first SGP after

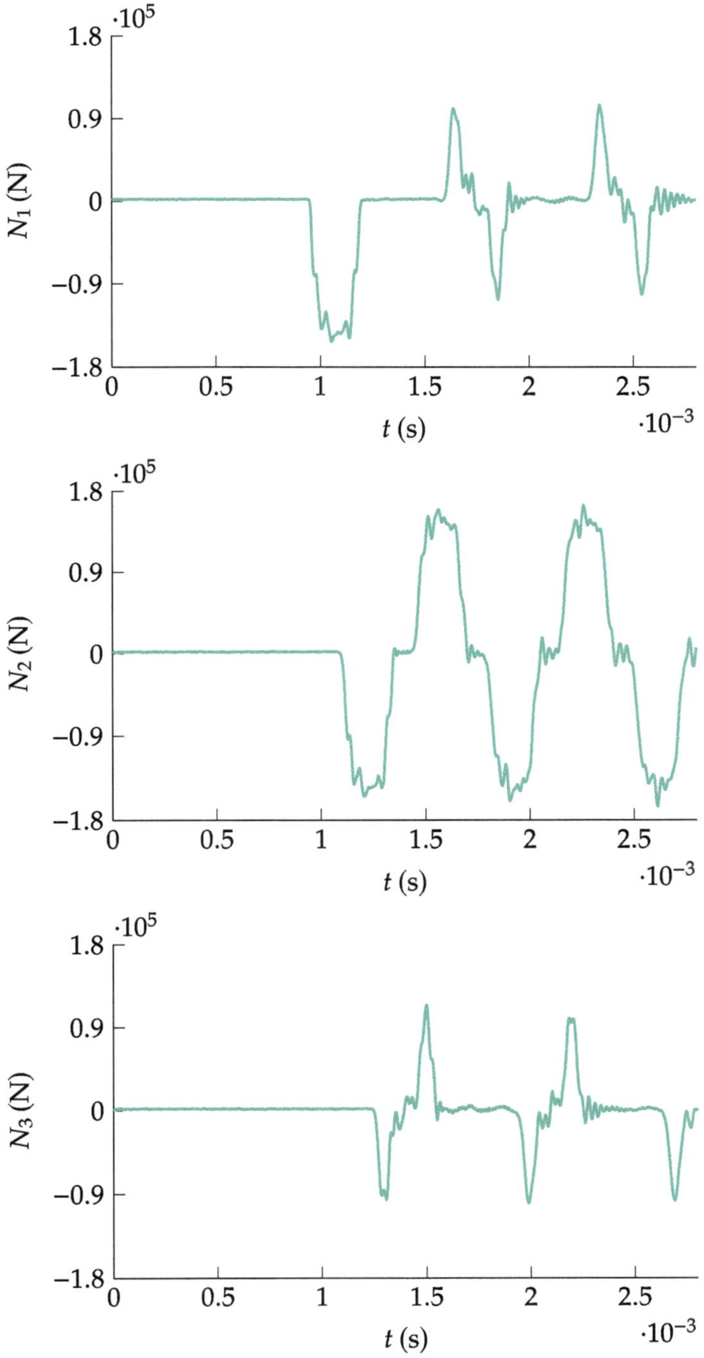

Figure 5.16: Forces versus time (second experiment) gained at the SGPs. The initial velocity of the piston is $4.58\,\mathrm{m\,s^{-1}}$.

the incident wave has passed with the difference that the sequence starts with a tensile peak.

Comparison with Numerical Results

The comparison of the experimental versus the simulation results at the first SGP is presented in Fig. 5.17a. For detailed analysis, the incident wave is magnified in the zoom box (filtered on the left side and unfiltered on the right side of Fig. 5.17b). It is only possible to generate filtered results for the simulation as unfiltered measurements would include aliasing effects. The final shape of the incident wave consists of a multitude of superimposed waves that are transmitted and reflected several times within the piston. However, some waves significantly contribute to the outcome. These waves are illustrated in Fig. 5.17c.

The qualitative and quantitative behavior matches very well with only small deviations between the results. The greatest qualitative deviation occurs approximately in the middle of the filtered results. At this position, the simulation result is shortly deflecting in negative direction while the experimental result is continuously increasing. Rod and piston do not separate when the reflected wave at the free end of the piston reaches the transition zone but shortly later which leads to a wider incident wave compared to the first experiment. It is very difficult to analyze the origin of every oscillation in the results as the longer the time proceeds the more piston variations influence the force shape. Moreover, the steps blur due to filtering. That is the reason why unfiltered results are used to analyze the most significant steps.

The first step at t_1 originates from the wave which propagates into the piston, reflects at the beginning of the cone and transmits back to the first SGP. From t_1 until close to t_2, the force decreases almost linearly which is due to the conical part of the piston. When the wave, which reflects at the end of the cone, reaches SGP_1 at t_2, the force rises immediately. The steps at t_3/t_4 refer to the waves that are reflected at the beginning/end of the middle block. The reflected wave at the small block close to the end of the piston is responsible for the jump in force at t_5. The biggest jump in force occurs at t_6. It originates from the wave that transmits through the piston, reflects at the piston's free end and transmits back to SGP_1.

While comparing the unfiltered with the filtered results, it is noticeable that not every detail is represented in the filtered result. However, the main changes are visible. The comparison shows that the main variations in the piston geometry are depicted in both

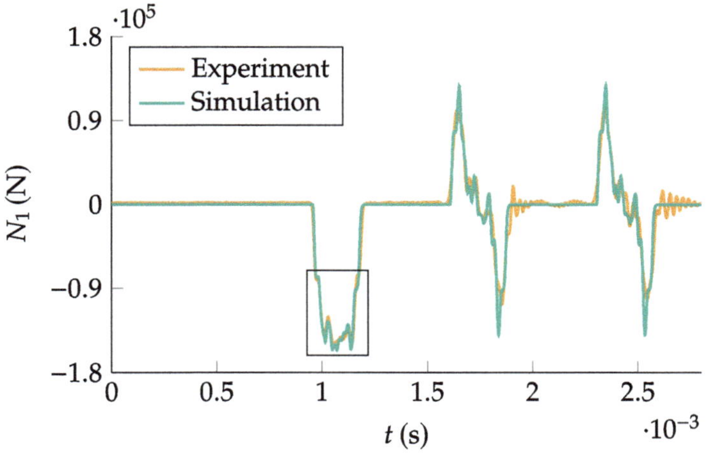

(a) Comparison between experimental and simulation results. The initial velocity of the piston is $4.58\,\mathrm{m\,s^{-1}}$.

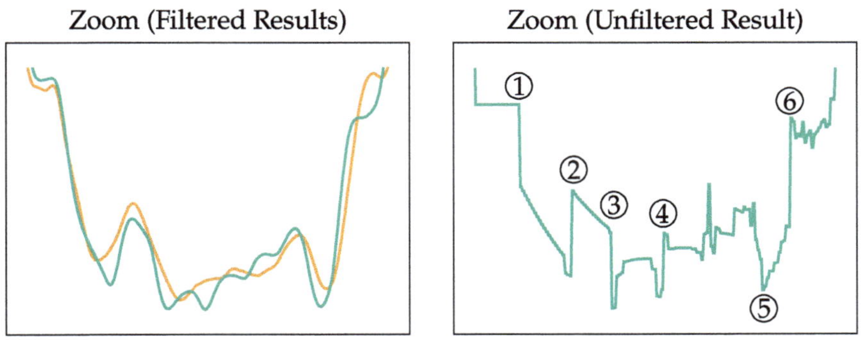

(b) Zoom of the framed section in (a) for filtered results (left) and the unfiltered simulation result (right).

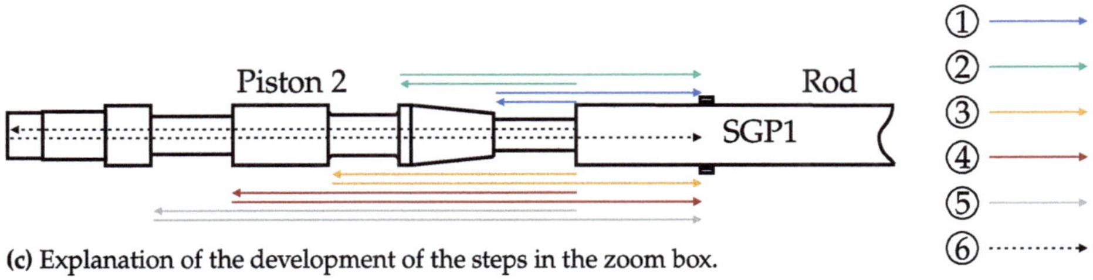

(c) Explanation of the development of the steps in the zoom box.

Figure 5.17: Evaluation of the second experiment.

experimental and simulation results with very good accordance. Therefore, it can be concluded that the 1D model is validated. Moreover, the comparisons suggest that for the given impact problems, the 1D consideration is sufficient. The advantage of the 1D theory is that the influence of the respective piston sections may be worked out. Furthermore, the calculation time is much shorter. The main advantage is, however, that it is possible to develop a method that solves the inverse problem based on the 1D theory.

6 Conclusions

The thesis aims to develop a method related to inverse problems in impacting rods. Specifically, the impedance of the impacting rod is determined so that a prescribed force at an arbitrary axial position of the second rod is generated (*impactor synthesis problem*). The method is relevant for several experimental and practical applications. For example, in percussion drilling, the question arises how the impedance of the impacting rod can be determined to generate an incident force that improves the efficiency of the drilling process.

The *impactor synthesis problem* is only addressed by a few researchers. The most promising approach solves the inverse problem in terms of an integral equation. By applying the integral based approach, several approximation methods are carried out to calculate the impactor's impedance. Moreover, it is only possible to prescribe the desired force at the hitting surface. In the context of the thesis, an approach based on the idea of differential methods is developed. It addresses the drawbacks of the integral based method mentioned above.

The method relies on the idea of directly exploiting the structure of the wave propagation problem. Therefore, the impedance functions of the rods are approximated by piecewise constant step functions. Use is made of the circumstance that each element of the unknown impactor is influencing the prescribed force at different moments for the first time. The element, which just influences the actual force for the first time, is adjusted so that the actual force matches the prescribed force. Thus, the elements are determined recursively starting from the element which is closest to the prescribed force position.

For piecewise constant impedance functions, the developed method delivers exact solutions in closed-form. Moreover, a condition is derived which states if a solution of the inverse problem exists. It is shown that any additional element, whose impedance differs from the previous one, diminishes the probability that a solution exists. The method allows to prescribe the force at any axial position of the second rod. Furthermore, arbitrary initial conditions are applicable to the second rod, including external forces.

The practical applicability of the method is validated in two steps. Firstly, the determined impedance function generates the prescribed force over time which is calculated by applying the 1D model. Secondly, the 1D model is validated experimentally by comparing experimental versus simulation results. In particular, it is shown that the model is also valid for impacting rods with complex geometry variations. As scattering problems are special cases of the *impactor synthesis problem*, the developed method is widely applicable.

List of Abbreviations

FEM	Finite element method
PDE	Partial differential equation
RM	Recursive method
SG	Strain gauge
SGP	Strain gauge position

List of Symbols

Indices

+	Acting on the right end of the element	
−	Acting on the left end of the element	
I	Impact	
dt	Dead time	
i	Incident	
in	Initial	
r	Reflected	
t	Transmitted	

Symbols

A	Cross section area	(m^2)
A_0	Cross section area at $x = 0$	(m^2)
B	Constant	(−)
C	Coefficient	(−)
E	Young's modulus	(N/m^2)
F^*	External force	(N)
G_j	Coefficients of the impulse response	(N)
$\mathcal{H}$	Heaviside function	(−)
I	Measurable quantity at the border of the scattering problem	(Case dependent)
$\mathcal{K}$	Linear differential operator	(−)
N	Cutting force, positive in tension	(N)
$\hat{N}$	Initial incident force acting at the rock	(N)

O	Measurable quantity at the border of the scattering problem	(Case dependent)
$\mathcal{O}$	Dimension of the computational cost	(–)
U	Eigenfunction	(m)
X_j	Coefficients of the piecewise constant force N_{I}	(N)
Y_j	Coefficients of the piecewise constant force N_p	(N)
Z	Impedance	(N s/m)
$\tilde{Z}$	Impedance of Rod 2	(N s/m)
a	Wave amplitude	(m)
a_0	Wave amplitude at $x = 0$	(m)
b	Summarized expression	(m^{-2})
c	Wave propagation speed	(m/s)
e	Ratio between maximum bending and normal stress	(%)
f	Arbitrary function	(–)
g	Arbitrary function	(–)
$\bar{g}$	Gravitational constant	(N/kg)
h	Summarized expression	(m^{-1})
k	Wave number	(m^{-1})
ℓ	Length	(m)
ℓ_{Pis}	Length of the piston	(m)
m	Mass of Rod 2	(kg)
n	Number of elements of Rod 1	(–)
$\tilde{n}$	Number of elements of Rod 2	(–)
$\bar{n}$	Number of time intervals	(–)
$\hat{n}$	Number of nodes in contact with bearings of Rod 2	(–)
n^*	Total number of rod elements	(–)
s	Distance from the hitting surface in direction of Rod 1	(m)
t	Time	(s)
t_0	Time constant	(s)
t^*	Time variable	(s)
u	Particle displacement	(m)
u_0	Particle displacement at $t = 0$	(m)
v	Particle velocity	(m/s)
v_0	Particle velocity at $t = 0$	(m/s)
v_{in}	Initial velocity of Rod 1	(m/s)
$\tilde{v}_{\mathrm{in}}$	Initial velocity of Rod 2	(m/s)
x	Spatial coordinate	(m)

α	Summarized expression	(m^{-1})
β	Summarized expression	(m^{-1})
δ_{jk}	Kronecker symbol	(−)
Δt	Time step	(s)
ε	Axial strain	(−)
γ	Distributed parameter of the scattering problem	(Case dependent)
ω	Angular frequency	(rad/s)
ω_k	kth eigenfrequency	(rad/s)
ρ	Mass density	(kg/m^3)
σ	Stress	(N/m^2)
ϑ	Summarized expression	(kg/m)
$\theta(s)$	Time it takes for the wave front to travel from the impact surface to position s	(s)
ξ	Constant	(m^{-1})
ζ	Distributed parameter of the scattering problem	(Case dependent)

Bibliography

[1] Abrate, S. (1995): Vibration of non-uniform rods and beams. *Journal of Sound and Vibration* **185**(4), pp. 703 – 716.

[2] Achenbach, J. D. (1973): *Wave propagation in elastic solids*. 1st ed. Vol. 16. North-Holland series in applied mathematics and mechanics. Amsterdam, New York, Oxford: North-Holland.

[3] Agranovich, Z. S. and V. A. Marchenko (1963): *The Inverse Problem of Scattering Theory*. Vol. 12. Russian Monographs and Texts on Advanced Mathematics and Physics. New York: Gorden and Breach.

[4] Aktosun, T. (2006): Inverse Scattering to Determine the Shape of a Vocal Tract. In: *The Extended Field of Operator Theory*. M. A. Dritschel ed. Vol. 171. Operator Theory Advances and Applications. Basel: Birkhäuser, pp. 1 – 16.

[5] Anderson, B. D. O. and T. Kailath (1978): Fast algorithms for the integral equations of the inverse scattering problem. *Integral Equations and Operator Theory* **1**(1), pp. 132 – 136.

[6] Bancroft, D. (1941): The velocity of longitudinal waves in cylindrical bars. *Physical Review* **59**(7), pp. 588 – 593.

[7] Bapat, C. (1995): Vibration of rods with uniformly tapered sections. *Journal of Sound and Vibration* **185**(1), pp. 185 – 189.

[8] Bardan, V. and E. A. Robinson (2018): Inverse problem for Goupillaud-layered earth model and dynamic deconvolution. *Geophysical Prospecting* **66**(8), pp. 1441 – 1456.

[9] Beccu, R. and B. Lundberg (1987): Transmission and dissipation of stress wave energy at a percussive drill rod joint. *International Journal of Impact Engineering* **6**(3), pp. 157 – 173.

[10] Berryman, J. G. and R. R. Greene (1980): Discrete inverse methods for elastic waves in layered media. *Geophysics* **45**(2), pp. 213 – 233.

[11] Bruckstein, A. M., B. C. Levy, and T. Kailath (1985): Differential methods in inverse scattering. *SIAM Journal on Applied Mathematics* **45**(2), pp. 312 – 335.

[12] Bruckstein, A. M. and T. Kailath (1987): Inverse scattering for discrete transmission-line models. *SIAM Review* **29**(3), pp. 359 – 389.

[13] Bu, C., Y. Qu, Z. Cheng, and B. Liu (2009): Numerical simulation of impact on pneumatic DTH hammer percussive drilling. *Journal of Earth Science* **20**(5), pp. 868 – 878.

[14] Bube, K. P. and R. Burridge (1983): The one-dimensional inverse problem of reflection seismology. *SIAM Review* **25**(4), pp. 497 – 559.

[15] Burkhart, S. and R. Wilcox (1990): Arbitrary pulse shape synthesis via nonuniform transmission lines. *IEEE Transactions on Microwave Theory and Techniques* **38**(10), pp. 1514 – 1518.

[16] Caltaru, M. M., R. G. Ripeanu, M. Badicioiu, and D. G. Zisopol (2020): Experimental researches to establish the optimum hardbanding technology and materials of the heavy weight drill pipe. Pnr 01017. *MATEC Web of Conferences* **318**. G. Mansour, D. Tzetzis, and P. Kyratsis eds.

[17] Carlsson, J., K. Sundin, and B. Lundberg (1990): A method for determination of in-hole dynamic force-penetration data from two-point strain measurement on a percussive drill rod. *International Journal of Rock Mechanics and Mining Sciences & Geomechanics Abstracts* **27**(6), pp. 553 – 558.

[18] Chiang, L. E. (2004): Dynamic force – penetration curves in rock by matching theoretical to experimental wave propagation response. *Experimental Mechanics* **44**(2), pp. 167 – 175.

[19] Chiang, L. E. and D. A. Elias (2008): A 3D FEM methodology for simulating the impact in rock-drilling hammers. *International Journal of Rock Mechanics and Mining Sciences* **45**(5), pp. 701 – 711.

[20] Choe, J. and H. C. Juvkam-Wold (2002): Pile driving analysis for top hammering and bottom hammering. *Journal of Geotechnical and Geoenvironmental Engineering* **128**(2), pp. 174 – 182.

[21] Chree, C. (1889): The equation of an isotropic elastic solid in polar and cylindrical coordinates, their solution and applications. *Transactions of the Cambridge Philosophical Society* **14**, pp. 250 – 369.

[22] Christensen, R. J., S. R. Swanson, and W. S. Brown (1972): Split-Hopkinson-Bar tests on rock under confining pressure. *Experimental Mechanics* **12**(11), pp. 508 – 513.

[23] Doyle, J. F. (1997): *Wave Propagation in Structures*. New York, Berlin, Heidelberg: Springer.

[24] Dutta, P. (1968): The determination of stress waveforms produced by percussive drill pistons of various geometrical designs. *International Journal of Rock Mechanics and Mining Sciences & Geomechanics Abstracts* **5**(6), pp. 501 – 510.

[25] FOLLANSBEE, P. S. and C. FRANTZ (1983): Wave propagation in the Split Hopkinson Pressure Bar. *Journal of Engineering Materials and Technology* **105**(1), pp. 61–66.

[26] FRANTZ, C., P. FOLLANSBEE, and W. WRIGHT (1984): *New Experimental Techniques with the Split Hopkinson Pressure Bar*. Tech. rep. LA-UR-83-3422, CONF-840653-1. Los Alamos National Lab.

[27] GEORGIOS, V., C. FOLCO, and S. GEORGE (2013): A numerical study of wave dispersion curves in cylindrical rods with circular cross-section. *Journal of Applied and Computational Mechanics* **7**(1), pp. 99–114.

[28] GERLACH, R., S. K. SATHIANATHAN, C. SIVIOUR, and N. PETRINIC (2011): A novel method for pulse shaping of Split Hopkinson Tensile Bar signals. *International Journal of Impact Engineering* **38**(12), pp. 976–980.

[29] GILAT, A., T. E. SCHMIDT, and A. L. WALKER (2008): Full field strain measurement in compression and Tensile Split Hopkinson Bar experiments. *Experimental Mechanics* **49**(2), pp. 291–302.

[30] GLADWELL, G. M. L. (2012): *Inverse Problems in Scattering*. Solid Mechanics and its Applications. Dordrecht, Boston, London: Kluwer Academic Publishers.

[31] GOLDSMITH, W. (1960): *Impact: The Theory and Physical Behaviour of Colliding Solids*. London: Arnold.

[32] GRAFF, K. F. (1991): *Wave Motion in Elastic Solids*. New York: Dover Publications.

[33] GRAY, G. T. (2000): Classic Split-Hopkinson Pressure Bar testing. In: *Mechanical Testing and Evaluation*. 10th ed. Metals Park, Ohio: ASM International, pp. 462–476.

[34] GUO, S. and S. YANG (2011): Wave motions in non-uniform one-dimensional waveguides. *Journal of Vibration and Control* **18**(1), pp. 92–100.

[35] HAGEDORN, P. (1989): *Technische Schwingungslehre*. 1st ed. Vol. 2. Lineare Schwingungen Kontinuierlicher Mechanischer Systeme. Berlin, Heidelberg: Springer.

[36] HAGEDORN, P. and A. DASGUPTA (2007): *Vibrations and Waves in Continuous in Continuous Mechanical Systems*. Chichester: John Wiley & Sons.

[37] JIMENO, E. L., A. CARCEDO, and F. J. AYALA-CARCEDO (1995): *Drilling and Blasting of Rocks*. London, New York: Taylor & Francis.

[38] JOHNSON, W. (1972): *Impact Strength of Materials*. London: Arnold.

[39] KARLSSON, L., B. LUNDBERG, and K. SUNDIN (1989): Experimental study of a percussive process for rock fragmentation. *International Journal of Rock Mechanics and Mining Sciences & Geomechanics Abstracts* **26**(1), pp. 45–50.

[40] KOLSKY, H. (1949): An investigation of the mechanical properties of materials at very high rates of loading. *Proceedings of the Physical Society. Section B* **62**(11), pp. 676–700.

[41] Kolsky, H. (1963): *Stress Waves in Solids*. Dover Books on Physics and Mathematical Physics. New York: Dover Publications.

[42] Krivtsov, A. M. and M. Wiercigroch (2000): Penetration rate prediction for percussive drilling via dry friction model. *Chaos, Solitons & Fractals* **11**(15), pp. 2479 – 2485.

[43] Kumar, B. and R. Sujith (1997): Exact solutions for the longitudinal vibration of non-uniform rods. *Journal of Sound and Vibration* **207**(5), pp. 721 – 729.

[44] Li, Q. (2000): Exact solutions for free longitudinal vibrations of non-uniform rods. *Journal of Sound and Vibration* **234**(1), pp. 1 – 19.

[45] Li, X., T. Lok, J. Zhao, and P. Zhao (2000): Oscillation elimination in the Hopkinson Bar apparatus and resultant complete dynamic stress – strain curves for rocks. *International Journal of Rock Mechanics and Mining Sciences* **37**(7), pp. 1055 – 1060.

[46] Lundberg, B. and M. Lesser (1977): Determination of the distributed parameters of a lossless wave supporting system from boundary measurements. *Applied Scientific Research* **33**(2), pp. 151 – 162.

[47] Lundberg, B. and M. Lesser (1978): On impactor synthesis. *Journal of Sound and Vibration* **58**(1), pp. 5 – 14.

[48] Lundberg, B. and M. Okrouhlik (2001): Influence of 3D effects on the efficiency of percussive rock drilling. *International Journal of Impact Engineering* **25**(4), pp. 345 – 360.

[49] Lundberg, B. and P. Collet (2010): Optimal wave with respect to efficiency in percussive drilling with integral drill steel. *International Journal of Impact Engineering* **37**(8), pp. 901 – 906.

[50] Mang, H. A. and G. Hofstetter (2018): *Festigkeitslehre*. 5th ed. Berlin: Springer.

[51] Mao, J.-F. and Z.-F. Li (1992): Analysis of the time response of nonuniform multiconductor transmission lines with a method of equivalent cascaded network chain. *IEEE Transactions on Microwave Theory and Techniques* **40**(5), pp. 948 – 954.

[52] Mao, J.-F., O. Wing, and F.-Y. Chang (1996): Transmission line synthesis by the method of characteristics. *IEEE Transactions on Circuits and Systems I: Fundamental Theory and Applications* **43**(6), pp. 461 – 468.

[53] Marais, S. T., R. B. Tait, T. J. Cloete, and G. N. Nurick (2004): Material testing at high strain rate using the Split Hopkinson Pressure Bar. *Latin American Journal of Solids and Structures* **1**, pp. 319 – 339.

[54] Mermelstein, P. (1967): Determination of the vocal-tract shape from measured formant frequencies. *The Journal of the Acoustical Society of America* **41**(5), pp. 1283 – 1294.

[55] MOHR, D., G. GARY, and B. LUNDBERG (2010): Evaluation of stress – strain curve estimates in dynamic experiments. *International Journal of Impact Engineering* **37**(2), pp. 161 – 169.

[56] ÖDEEN, S. and B. LUNDBERG (1991): Prediction of impact force by impulse response method. *International Journal of Impact Engineering* **11**(2), pp. 149 – 158.

[57] PEIRS, J., P. VERLEYSEN, W. V. PAEPEGEM, and J. DEGRIECK (2011): Determining the stress – strain behaviour at large strains from high strain rate tensile and shear experiments. *International Journal of Impact Engineering* **38**(5), pp. 406 – 415.

[58] POCHHAMMER, L. (1876): Über die Fortpflanzungsgeschwindigkeiten kleiner Schwingungen in unbegrenzten isotropen Kreiszylinder (in German). *Zeitschrift für Reine und Angewandte Mathematik* **81**, pp. 324 – 336.

[59] RAJ, A. and R. SUJITH (2005): Closed-form solutions for the free longitudinal vibration of inhomogeneous rods. *Journal of Sound and Vibration* **283**, pp. 1015 – 1030.

[60] ROBINSON, E. A. (1975): Dynamic predictive deconvolution. *Geophysical Prospecting* **23**(4), pp. 779 – 797.

[61] ROBIT (2020): *Top Hammer Drilling*. Visited on 31.03.2020. URL: `https://www.robitgroup.com/products/top-hammer-drilling/`.

[62] SANTOSA, F. and H. SCHWETLICK (1982): The inversion of acoustical impedance profile by methods of characteristics. *Wave Motion* **4**(1), pp. 99 – 110.

[63] SARWAR, A. K. M. and A. J. RUDMAN (1985): Inversion of a normally incident reflection seismogram by the Gopinath – Sondhi integral equation. *Geophysical Journal International* **81**(3), pp. 551 – 562.

[64] SARWAR, A. and D. SMITH (1987): Wave scattering deconvolution by seismic inversion. *Geophysical Prospecting* **35**(5), pp. 491 – 501.

[65] SCHROETER, J. and M. M. SONDHI (1994): Techniques for estimating vocal-tract shapes from the speech signal. *IEEE Transactions on Speech and Audio Processing* **2**(1), pp. 133 – 150.

[66] SHORR, B. (1995): Analysis of wave propagation in elastic-plastic rods of a variable cross section using direct mathematical modelling. *Archive of Applied Mechanics* **65**(8), pp. 537 – 547.

[67] SHORR, B. F. (2003): *The Wave Finite Element Method*. V. I. Babitsky and J. Wittenburg eds. 1st ed. Foundations of Engineering Mechanics. Berlin, Heidelberg: Springer.

[68] SONDHI, M. M. and B. GOPINATH (1971): Determination of vocal-tract shape from impulse response at the lips. *The Journal of the Acoustical Society of America* **49**(6B), pp. 1867 – 1873.

[69] Sondhi, M. M. and J. R. Resnick (1983): The inverse problem for the vocal tract: Numerical methods, acoustical experiments, and speech synthesis. *The Journal of the Acoustical Society of America* **73**(3), pp. 985 – 1002.

[70] Sondhi, M. M. (1984): A Survey of the Vocal Tract Inverse Problem: Theory, Computations and Experiments. In: F. Santosa, Y. H. Pao, W. W. Symes, and C. Holland eds. Inverse Problems of Acoustic and Elastic Waves. Philadelphia: SIAM, pp. 1 – 19.

[71] Symes, W. W. (1981): Stable solution of the inverse reflection problem for a smoothly stratified elastic medium. *SIAM Journal on Mathematical Analysis* **12**(3), pp. 421 – 453.

[72] Wittenburg, J. (2013): *Schwingungslehre*. Berlin, Heidelberg: Springer.

[73] Xia, K. and W. Yao (2015): Dynamic rock tests using Split Hopkinson (Kolsky) Bar system – A review. *Journal of Rock Mechanics and Geotechnical Engineering* **7**(1), pp. 27 – 59.

[74] Zakharov, V. E. and A. B. Shabat (1972): Exact theory of two-dimensional self-focusing and one-dimensional self-modulation of waves in nonlinear media. *Soviet Physics JETP* **34**(1), pp. 62 – 69.

[75] Zhou, Y. X., K. Xia, X. B. Li, H. B. Li, G. W. Ma, J. Zhao, Z. L. Zhou, and F. Dai (2011): Suggested Methods for Determining the Dynamic Strength Parameters and Mode-I Fracture Toughness of rock Materials. In: *The ISRM Suggested Methods for Rock Characterization, Testing and Monitoring: 2007-2014*. R. Ulusay ed. Cham, Heidelberg: Springer, pp. 35 – 44.

Publications

[76] Burgert, J., W. Seemann, and F. Hartenbach (2017): Optimization of the piston geometry to generate a desired stress wave shape. *PAMM* **17**(1), pp. 675 – 676.

[77] Burgert, J. and W. Seemann (2018): Comparison of a numerical versus an analytical approach to calculate 1D impact problems of non-uniform cross sections. Pnr e201800250. *PAMM* **18**(1).

[78] Burgert, J., W. Seemann, C. Moilanen, J. Montonen, J. Miettinen, and P. Saarenrinne (2018): Pneumatic single hit test rig for analysis of the stress wave profile of non-uniform impacting rods. In: *Proceedings of ISMA2018, International Conference on Noise and Vibration Engineering/USD2018*. W. Desmet, B. Pluymers, D. Moens, and W. Rottiers eds. Leuven: KU Leuven, pp. 3535 – 3547.

[79] Burgert, J. and W. Seemann (2019): Measuring the wave shape and wave propagation speed of two impacting rods of arbitrary cross section and material. In: *Proceedings of the 26th International Congress on Sound and Vibration*. Montreal: Canadian Acoustical Association, pp. 199 – 206.

[80] Burgert, J. and W. Seemann (2019): Numerical method to determine the inverse solution of two impacting rods of non-constant cross section. Pnr e201900007. *PAMM* **19**(1).

KARLSRUHER INSTITUT FÜR TECHNOLOGIE (KIT)
SCHRIFTENREIHE DES INSTITUTS FÜR TECHNISCHE MECHANIK (ITM)

ISSN 1614-3914

Band 1 **Marcus Simon**
Zur Stabilität dynamischer Systeme mit stochastischer Anregung. 2004
ISBN 3-937300-13-9

Band 2 **Clemens Reitze**
Closed Loop, Entwicklungsplattform für mechatronische Fahrdynamikregelsysteme. 2004
ISBN 3-937300-19-8

Band 3 **Martin Georg Cichon**
Zum Einfluß stochastischer Anregungen auf mechanische Systeme. 2006
ISBN 3-86644-003-0

Band 4 **Rainer Keppler**
Zur Modellierung und Simulation von Mehrkörpersystemen unter Berücksichtigung von Greifkontakt bei Robotern. 2007
ISBN 978-3-86644-092-0

Band 5 **Bernd Waltersberger**
Strukturdynamik mit ein- und zweiseitigen Bindungen aufgrund reibungsbehafteter Kontakte. 2007
ISBN 978-3-86644-153-8

Band 6 **Rüdiger Benz**
Fahrzeugsimulation zur Zuverlässigkeitsabsicherung von karosseriefesten Kfz-Komponenten. 2008
ISBN 978-3-86644-197-2

Band 7 **Pierre Barthels**
Zur Modellierung, dynamischen Simulation und Schwingungsunterdrückung bei nichtglatten, zeitvarianten Balkensystemen. 2008
ISBN 978-3-86644-217-7

Die Bände sind unter www.ksp.kit.edu als PDF frei verfügbar oder als Druckausgabe zu bestellen.

Band 8 **Hartmut Hetzler**
Zur Stabilität von Systemen bewegter Kontinua mit Reibkontakten am Beispiel des Bremsenquietschens. 2008
ISBN 978-3-86644-229-0

Band 9 **Frank Dienerowitz**
Der Helixaktor – Zum Konzept eines vorverwundenen Biegeaktors. 2008
ISBN 978-3-86644-232-0

Band 10 **Christian Rudolf**
Piezoelektrische Self-sensing-Aktoren zur Korrektur statischer Verlagerungen. 2008
ISBN 978-3-86644-267-2

Band 11 **Günther Stelzner**
Zur Modellierung und Simulation biomechanischer Mehrkörpersysteme. 2009
ISBN 978-3-86644-340-2

Band 12 **Christian Wetzel**
Zur probabilistischen Betrachtung von Schienen- und Kraftfahrzeugsystemen unter zufälliger Windanregung. 2010
ISBN 978-3-86644-444-7

Band 13 **Wolfgang Stamm**
Modellierung und Simulation von Mehrkörpersystemen mit flächigen Reibkontakten. 2011
ISBN 978-3-86644-605-2

Band 14 **Felix Fritz**
Modellierung von Wälzlagern als generische Maschinenelemente einer Mehrkörpersimulation. 2011
ISBN 978-3-86644-667-0

Band 15 **Aydin Boyaci**
Zum Stabilitäts- und Bifurkationsverhalten hochtouriger Rotoren in Gleitlagern. 2012
ISBN 978-3-86644-780-6

Band 16 **Rugerri Toni Liong**
Application of the cohesive zone model to the analysis of rotors with a transverse crack. 2012
ISBN 978-3-86644-791-2

Band 17 **Ulrich Bittner**
Strukturakustische Optimierung von Axialkolbeneinheiten. Modellbildung, Validierung und Topologieoptimierung. 2013
ISBN 978-3-86644-938-1

Band 18 **Alexander Karmazin**
Time-efficient Simulation of Surface-excited Guided Lamb Wave Propagation in Composites. 2013
ISBN 978-3-86644-935-0

Band 19 **Heike Vogt**
Zum Einfluss von Fahrzeug- und Straßenparametern auf die Ausbildung von Straßenunebenheiten. 2013
ISBN 978-3-7315-0023-0

Band 20 **Laurent Ineichen**
Konzeptvergleich zur Bekämpfung der Torsionsschwingungen im Antriebsstrang eines Kraftfahrzeugs. 2013
ISBN 978-3-7315-0030-8

Band 21 **Sietze van Buuren**
Modeling and simulation of porous journal bearings in multibody systems. 2013
ISBN 978-3-7315-0084-1

Band 22 **Dominik Kern**
Neuartige Drehgelenke für reibungsarme Mechanismen. 2013
ISBN 978-3-7315-0103-9

Band 23 **Nicole Gaus**
Zur Ermittlung eines stochastischen Reibwerts und dessen Einfluss auf reibungserregte Schwingungen. 2013
ISBN 978-3-7315-0118-3

Band 24 **Fabian Bauer**
Optimierung der Energieeffizienz zweibeiniger Roboter durch elastische Kopplungen. 2014
ISBN 978-3-7315-0256-2

Die Bände sind unter www.ksp.kit.edu als PDF frei verfügbar oder als Druckausgabe zu bestellen.

Band 25 **Benedikt Wiegert**
Nichtlineare Schwingungen von Systemen
mit elastohydrodynamischen Linienkontakten. 2015
ISBN 978-3-7315-0350-7

Band 26 **Arsenty Tikhomolov**
Analytische, numerische und messtechnische
Untersuchung der Dynamik von Fahrzeugkupplungen
am Beispiel des Trennproblems. 2015
ISBN 978-3-7315-0362-0

Band 27 **Daniel Maier**
On the Use of Model Order Reduction Techniques
for the Elastohydrodynamic Contact Problem. 2015
ISBN 978-3-7315-0369-9

Band 28 **Xiaoyu Zhang**
Crosswind stability of vehicles under nonstationary
wind excitation. 2015
ISBN 978-3-7315-0376-7

Band 29 **Jens Deppler**
Ein Beitrag zur viskoelastischen Modellierung nichtholonomer
Bindungsgleichungen. 2017
ISBN 978-3-7315-0548-8

Band 30 **Georg Jehle**
Zur Modellbildung und Simulation reibungserregter
Schwingungen in Pkw-Schaltgetrieben. 2017
ISBN 978-3-7315-0668-3

Band 31 **Joachim Klima**
Lubricant transport towards tribocontact in capillary
surface structures. 2018
ISBN 978-3-7315-0814-4

Band 32 **Gábor Kenderi**
Nonparametric identification of nonlinear dynamic systems. 2018
ISBN 978-3-7315-0834-2

Band 33 **Jan Henrik Schmidt**
An efficient solution procedure for elastohydrodynamic contact problems considering structural dynamics. 2018
ISBN 978-3-7315-0872-4

Band 34 **Ulrich Johannes Römer**
Über den Einfluss der Fußgeometrie auf die Energieeffizienz beim zweibeinigen Gehen. 2019
ISBN 978-3-7315-0887-8

Band 35 **Simon Kapelke**
Zur Beeinflussung reibungsbehafteter Systeme mithilfe überlagerter Schwingungen. 2019
ISBN 978-3-7315-0905-9

Band 36 **Jens Burgert**
On direct and inverse problems related to longitudinal impact of non-uniform elastic rods. 2021
ISBN 978-3-7315-1087-1